Moving Objects Management

Xiaofeng Meng • Zhiming Ding • Jiajie Xu

Moving Objects Management

Models, Techniques and Applications

Second Edition

With 105 Figures

Xiaofeng Meng
Renmin University of China
Beijing, China

Zhiming Ding
Jiajie Xu
Chinese Academy of Sciences
Beijing, China

ISBN 978-3-662-50562-5   ISBN 978-3-642-38276-5 (eBook)
DOI 10.1007/978-3-642-38276-5
Springer Heidelberg New York Dordrecht London

Jointly published with Tsinghua University Press, Beijing
Tsinghua University Press, Beijing

Preface

The widespread use of mobile positioning tools like GPS and smart mobile phones nowadays has aroused great interests in location-based services (LBS) that have to store and manage continuously changing positions of moving objects. This book gives a comprehensive and complete view of a moving objects database and introduces how it is used in LBS and transportation applications. It aims at moving objects management, from the location management perspective to analyze how the continually changing locations affect the traditional database and data mining technology. Specifically, the book describes the cutting edge technologies related to topics like moving objects modeling and location tracking, indexing and querying, trajectory prediction, location uncertainty, traffic flow analysis, objects clustering, traffic aware navigation and privacy issues as well as their application to intelligent transportation systems.

Previous studies mostly focused on moving objects database in free space. They assumed that the movement of the objects is unconstrained and based on Euclidean spaces. However, in the real world, objects usually move within spatially constrained networks, e.g., vehicles move on road networks. Overlooking this reality often leads to unrealistic data modeling and inaccurate query results. The content in this book focuses mainly on the moving objects within spatial networks, which is more practical. By exploiting the network feature of spatial networks, this book introduces models, techniques, and applications of moving objects management in a spatial network.

This book is intended to help readers understand the main technologies in moving object management and apply them to LBS and transportation applications. Compared with the first edition, this book particularly focuses on the constrained network environments, and it has made substantial changes to each chapter so that the cutting edge techniques in this field are included. With its accessible style and emphasis on practicality, the book presents new concepts and techniques for managing continuously moving objects. Database management systems developers, mobile applications developers, and applied R&D researchers will find the study an essential companion for new concepts, development strategies, and application models associated with this kind of changing location data. The book:

- Presents a comprehensive architecture of moving object management, which includes not only basic theories and new concepts but also practical technologies and applications
- Describes a set of new database techniques in modeling, tracking, indexing, querying of moving objects, traffic flow analysis, as well as data mining techniques in clustering analysis of moving objects
- Introduces some new research issues in location privacy and uncertainty management of moving objects, which are topics of major interest in this field
- Provides typical applications of moving objects management in intelligent transportation systems

Organization of the Book

This book contains 12 chapters, which describe the problems, models, techniques, and applications of moving objects management. It is organized as follows:

In Chap. 1, we introduce some background of moving objects management, including its concept and applications. Finally we present the main content: key technologies of moving objects databases and our focus in this book.

In Chap. 2, we introduce some underlying modeling methods and present two moving object models that can reflect real-time traffic conditions of the road network. The first one is the DTNMOM, which considers the dynamics of underlying road network. And for the second model called ARS-DTNMOM, we introduce the concept of atomic route section and define its corresponding data types and operations in database.

In Chap. 3, we introduce a few underlying methods on moving object tracking. Then, we describe three representative network-constrained location update strategies (Net-LUM, ANLUM, and EuNetMOD), which can achieve better performances in terms of communication costs and location tracking accuracy.

In Chap. 4, we first introduce a few of the underlying spatial index structures including the R-tree, TPR-tree, spatio-temporal R-tree, trajectory-bundle tree, and MON-tree. Then, we propose two new index methods that are used for indexing frequently updated trajectories of network constrained moving objects and indexing the whole trajectories with historical, current, and near future positions, respectively.

In Chap. 5, we classify the basic querying types for moving objects according to spatial predicates, temporal predicates, and moving spaces. Then, we introduce how to process a range query and a kNN query in a spatial network, based on the Euclidean restriction and network expansion frameworks.

In Chap. 6, we introduce advanced querying for moving objects including similar trajectory queries and density queries for moving objects in a spatial network. We first present how to process the snapshot density queries. Then, we introduce some efficient methods based on the safe interval to continuously monitor dense regions for moving objects.

In Chap. 7, we first review some linear prediction methods and analyze their limitations in handling moving objects in spatial networks, then present the simulation-based prediction methods: fast-slow bounds prediction and time-segment prediction, and finally present an uncertain path prediction method which can predict future trajectories based on the uncertain historic trajectories of moving objects in spatial networks.

In Chap. 8, we study the uncertainty management problem for moving objects databases with a few uncertainty models. Then we introduce a novel framework that can manage uncertainty trajectory effectively and answer queries about them accurately; particularly, we focus on the key technical issues like uncertain trajectory modeling, database operations, and query processing of uncertainty management.

In Chap. 9, we study the underlying researches and inherent problems in traffic behavior analysis based on moving object trajectories. Then we firstly propose a new model for objects moving on dynamic transportation networks (MODTN), based on which we introduce a real-time traffic flow statistical analysis method (NMOD-TFSA).

In Chap. 10, we introduce the clustering analysis of moving objects in spatial networks. After that, we introduce two new static clustering algorithms, which use the information of nodes and edges in the network to improve the clustering efficiency and accuracy. Then, we introduce the notion of cluster block (CB) as the underlying clustering unit and propose a unified framework of clustering moving objects in spatial network (CMON), which improves the dynamic clustering performance of moving objects and supports different clustering criteria. Finally, we introduce two trajectory clustering algorithms which use the partition-and-group framework for clustering trajectories and a filter-refinement framework for hot region discovery, respectively.

In Chap. 11, we present another application, traffic aware route navigation, with a new traffic aware route planning model based on incremental planning method introduced. By selecting intermediate destinations, a partial path rather than whole path is planned each time for long distance queries. In this way, route planning is more efficient because it is carried out in a much smaller region, and unnecessary re-calculations caused by the dynamic road conditions can be avoided.

In Chap. 12, we introduce location privacy, and analyze the challenges of preserving location. Then, we provide an analysis of the current studies including the system architecture, location anonymity, and query processing.

As shown in Fig. 1, the contents of the whole book construct a comprehensive moving object management and application system. Figure 1 also shows the relationship of each component in the system.

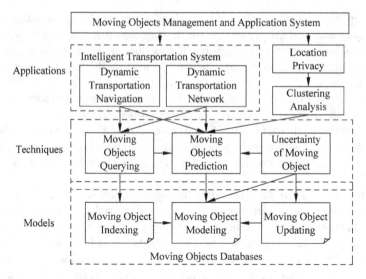

Fig. 1 Organization of the book

Acknowledgments

The work described in this book has been supported by the grants from the Natural Science Foundation of China (No. 61379050, 91024032, 91124001, 91224008); the National 863 High-tech Program (No. 2013AA013204); and Specialized Research Fund for the Doctoral Program of Higher Education (No. 20130004130001).

This book is based on the research work of the authors for over 15 years. The book integrates the collective intelligence from the mobile group of the WAMDM Lab (Lab of Web and Mobile Data Management) at Renmin University of China, and the database group of NFS Center (The National Engineering Research Center of Fundamental Software) at Institute of Software, Chinese Academy of Sciences. The authors would like to express their great thanks to all the people who contributed to this book, including Dr. Jidong Chen, Dr. Xiao Pan, Dr. Limin Guo, Dr. Kuien Liu, Dr. Haoming Guo, Xing Hao, Zhen Xiao, and Rui Ding. In particular, the authors wish to thank Dr. Jidong Chen, Dr. Xiao Pan, and Dr. Limin Guo for their valuable efforts on this book.

Beijing, China   Xiaofeng Meng
December 2012   Zhiming Ding
                                                                              Jiajie Xu

Contents

Acronyms

ADT	Abstract data type
ANN	Aggregate nearest neighbor
AU	Adaptive unit
CA	Cellular automaton
CN	Cluster node
CU	Cluster unit
DS	Dense segment
DSS	Dense segment set
DTTLU	Distance-threshold triggered location update
DyNSA	Dynamic navigation system based on moving objects stream aggregation
GCA	Graph of cellular automata
GPS	Global positioning system
HAT	Hierarchy aggregation tree
IER	Incremental Euclidean restriction
INE	Incremental network expansion
ITLU	ID-triggered location update
LBS	Location-based service
LP	Linear prediction
MBR	Minimum bounding rectangle
MO	Moving object
MOD	Moving objects databases
MODTN	Moving objects on dynamic transportation networks
MOST	Moving objects spatio-temporal
MRM	Mobile resource management
NN	Nearest neighbor
PDQ	Period density queries
PTSS	Prediction with time-segmented
QoS	Quality of service
RER	Range Euclidean restriction
RNE	Range network expansion

RNN	Reverse nearest neighbor
SDQ	Snap-shot density queries
SP	Simulation-based prediction
STTLU	Speed-threshold triggered location update
UT-Unit	Uncertain trajectory unit
UTR-Tree	Uncertain trajectory R-tree

Chapter 1
Introduction

Abstract The fast development of geo-positioning and wireless sensor network technologies has aroused widespread use of location-based services (LBS), which provide useful location-dependent information to users. LBS have become so important nowadays that people rely on it to plan trip, book cabs, and find share car partners. Moving objects database, which plays a key role in supporting LBS applications, has attracted great attention from both academy and industry in recent years. In this chapter, we introduce the concept of moving object data management first and then describe the wide applications of location-based service. Key techniques related to moving objects database are discussed and analyzed afterwards. After that, we mention the purpose and organization of this book.

Keywords Mobile computing • Location-based service • Moving objects management • Moving object databases • Model • Index • Query • Update • Prediction • Uncertainty management • Clustering • Traffic statistical analysis • Traffic navigation • Location privacy

1.1 Concept of Moving Objects Data Management

The general idea of moving object data management is to represent the moving entities in databases and process queries about them efficiently. Moving entities could be human, animals, all kinds of vehicles like cars, trucks, air planes, ships, etc., and people often issue queries about their location, such as finding all vacant taxicabs inside a requested spatial area. However, existing database management systems (DBMSs) are not well equipped to handle the massive dynamic location data sampled from moving objects. Therefore, moving objects database (MOD), which particularly includes the management of the moving objects location and related information, has become an enabling technology that can find various LBS applications nowadays.

X. Meng et al., *Moving Objects Management: Models, Techniques and Applications*, DOI 10.1007/978-3-642-38276-5__1,
© Tsinghua University Press, Beijing and Springer-Verlag Berlin Heidelberg 2014

Moving objects database belongs to the area of spatio-temporal databases, which in turn have its root in spatial database, dealing with descriptions of geometry in databases, and temporal database, addressing the development of data over time. The major difference between them is that moving objects database focuses on the continuous spatial position change with time (their movement is seen as trajectory), while other spatio-temporal databases only support the discrete changing of spatial information for all moving entities in database.

We can actually understand the idea of moving objects database from two different perspectives. Firstly, moving objects database is to represent, store, index, and query on the continuously changing locations of moving objects and to predict the future positions of them; secondly, the focus is to store the whole history of moving object movement in database, so as to answer queries on the location of moving objects at any instance (including both history and future). Essentially, the former approach is to analyze from the location management perspective, while the second one stands on the spatio-temporal data perspective. As a result, research on moving objects database can be made from two aspects as well: location management view and spatio-temporal data view.

The key problem of MOD is how to manage the locations of a set of moving objects in database, e.g., position of all taxicabs inside a city road network. Given a time instance, it is not a problem. However, as the taxicabs move, it is necessary to have the location frequently updated, so that we can derive its current location. Here we encounter an unpleasant trade-off between update cost and location precision. From the data management perspective, MOD focuses on issues like how to manage the information of moving objects location dynamically and how to process different types of complex queries about current and future positions efficiently.

So far, considerable research has been carried out on moving object data management. In the following sections, we introduce some typical applications and key technologies related to MOD, including the modeling and tracking of location information, spatio-temporal indexing, uncertainty management, query processing, trajectory data mining (including traffic flow analysis), and privacy issues.

1.2   Applications of Moving Objects Database

Moving objects database is a fundamental technique for LBS, from which people can get useful information and entertainment services through mobile devices. In a typical LBS application, moving objects use e-services that involve location information. The objects disclose their positional information (position, speed, velocity, etc.) to the services, which in turn use this and other information to provide specific functionality. The following five categories described next characterize what may be thought of as standard location-based services; they do not attempt to describe the diversity of services possible [28].

1. Traffic coordination and management: Based on past and up-to-date positional data on the subscribers to a service, the service may identify traffic jams and determine the currently fastest route between two positions; it may give estimates and accurate error bounds for the total travel time, and it may suggest updated routes for the remaining travel. It also becomes possible to automatically charge fees for the use of infrastructure such as highways or bridges (termed as road pricing and metered services).

2. Location-aware advertising and general content delivery: Users may receive sales information (or other content) based on their current locations when they indicate to the service that they are in "shopping mode." Positional data is used together with an accumulated user profile to provide a better service, e.g., advertisements that are more relevant to the user.

3. Integrated tourist services: This covers the advertising of the available options for various tourist services, including all relevant information about these services and options. Services may include overnight accommodation at campgrounds, hostels, and hotels; transportation via train, bus, taxi, or ferry; and cultural events, including exhibitions, concerts, etc. For example, this latter kind of service may cover opening-hour information, availability information, travel directions, directions to empty parking, and ticketing. It is also possible to give guided tours to tourists, e.g., that carry online cameras.

4. Safety-related services: It is possible to monitor tourists traveling in dangerous terrain and then react to emergencies (e.g., skiing or sailing accidents); it is possible to offer senile senior citizens more freedom of movement and a service that takes traffic conditions into account to guide users to desired destinations along safe paths.

5. Location-based games and entertainment: One example of this is treasure hunting, where the participants compete in recovering a treasure. The treasure is virtual, but is associated with a physical location. By monitoring the positions of the participants, the system is able to determine when the treasure is found and by whom. In a variation of this example, the treasure is replaced by a "monster" with "vision," "intelligence," and the ability to move. Another example in this category is a location-based ICQ service.

1.3   Key Technologies in Moving Objects Database

1.3.1   Moving Objects Modeling

The modeling of moving object in databases is a basic technique for MOD. In conventional databases, attribute values stored in table are assumed to be constant unless they are explicitly updated. However, in MOD, the location of moving object changes continuously, and people issue queries to find their history, current, and even future position. As conventional database models are unable to represent dynamic location information, moving objects modeling plays an important role in effective location management. Current research on MOD modeling can be generally classified into two categories: Euclidean (EU)-based modeling and network (NET)-based modeling.

EU-based modeling targets to represent the trajectories of free movement objects in Euclidean space. Wolfson et al. propose a moving objects spatio-temporal (MOST) model in [43, 51] first. Its core idea is to consider location of moving objects as a dynamic attribute, which is represented as a function of time. In this way, we do not need to update this attribute until this function is no longer valid. However, long trajectories cannot be well supported by the MOST model because of the limited representation capacity of simple functions. Later, Forlizzi et al. in [19] present a discrete moving object data model to overcome this drawback, with a feasible solution for complex moving object trajectory representation. Also, models like linear constraint [45], abstract data types [25], and space-time grid storage [11] are proposed. However, these EU-based models do not take into account the network constraint, while objects move with road network constraint in most real-life applications, especially for the vehicles in transportation scenarios.

As moving objects move according to the topology of underlying road network, the interaction between modeling and network structure enables better object movement representation, and this contributes to improve the performance of object tracking, data indexing, and query processing. To represent the movement of objects under road-network constraint, we need to model the road network first and then to model how the objects move on this network. Static road network can be generally represented in three ways: road-based representation, two-dimensional geographic coordinate-based representation, and graph-based representation. For the moving objects on road network, we can further use road segment and trajectory to model the movement of objects.

Vazirgiannis et al. propose a road-network-based moving object model that combines the trajectory with road network in [49], where road network is represented as a digital map and trajectory is represented as the path from starting point to destination. In recent years, some other network constraint models based on graph representation such as [24, 37, 44] have been proposed. But they simply assume linear movement and cannot reflect the real movement feature of moving objects in a road network, and they only consider static transportation networks. That limits their applicability in a majority of real-life applications. An advanced model is also proposed in [46] to integrate the past movement features to improve the capability of moving object representation.

1.3.2   Location Tracking of Moving Objects

In moving objects tracking, current position is periodically sent to the central server and stored in database. When the number of tracked moving objects becomes large, the scale of sampled data would be extremely huge if the update is too frequent. Therefore, a key issue is to find a proper balance between reducing update cost and improving precise location for query results. Current researches on moving object tracking mainly focus on location update and prediction, and they can be classified into two categories: EU-based tracking and NET-based tracking.

Existing EU-based tracking approaches generally include Fixed-Time Location Update Mechanisms (FTLU), Fixed-Distance Location Update Mechanisms (FDLU), and Euclidean-Motion-Vector-Based Location Update Mechanisms (Eu-MVLU). FTLU and FDLU mean to update according to fixed time duration and Euclidean distance, respectively, and they are widely used in real-world systems because of their simplicity. Eu-MVLU assumes the movements of an object can be represented with a motion vector. Eu-MVLU is generally much superior to FTLU and FDLU in terms of accuracy, and as a result, it has become increasingly influential in moving objects databases.

As for the road-network-based tracking, existing update strategies are mainly based on Network-Motion-Vector-Based Location Update Mechanism (Net-MVLU), and they can be classified according to the threshold, future position prediction, and group-based update mode. In threshold-based Net-MVLU approaches, we only update location when threshold is reached, so as to reduce update cost. Wolfson et al. propose a "dead-reckoning" tracking strategy in [50] first, but it only works for the scenarios with fixed and known path. Later, people improve it by proposing a new dead-reckoning policy based on angular and linear deviation [22]. Lam et al. further develop an adaptive threshold-based update mechanism that considers the effect of continuous queries on threshold [34], with a core idea to improve the accuracy of location tracking on objects (by setting lower threshold) frequently covered by query results.

More recently, update mechanisms [12, 13, 52] based on location prediction have been proposed to improve tracking accuracy. They use different functions (e.g., linear and constant functions) to estimate future location of moving objects and update only if the difference between sampled and estimated locations exceeds a threshold. As their tracking performance greatly depends on estimating function, Civilis et al. in [12, 13] propose three updated policies: point, vector, and segment-based policies. Furthermore, Ding et al. discuss the use of what is essentially segment-based tracking in [16] on the basis of their proposed data model for the management of road-network-constrained moving objects [14].

Different from most existing update strategies that aim to improve the efficiency of single object tracking, group-based update strategy [10] uses clustering techniques to reduce the communication cost. Specifically, moving objects are clustered into groups based on position and velocity, and location tracking is carried out on the group level. Repeated location data upload from objects in a same cluster can be avoided as a result.

1.3.3   Moving Objects Database Indexes

As the data sampled from moving objects could be extremely large, another key problem is how to construct proper indexes for speeding up query processing. Conventional spatio-temporal indexes fail here because of the high dynamics of moving objects, which lead to the frequent updates of indexes and then cause huge overloads. So far, many moving objects index structures have been proposed to handle this issue, and they can be generally classified to three categories: (1) historical trajectory-based indexes, (2) current location-based indexes, and (3) future position-based indexes.

Trajectory data are usually scalable dynamic data. Several indexing approaches [36, 40, 46] based on 3D variations of R-tree and R*-tree have been proposed to manage them, with a goal to minimize storage and query cost. However, the structure of underlying network is not considered in these indexes. In recent years, more efforts have been made for trajectory data indexing on moving objects in spatial networks, with typical methods like dimension transformation-based indexes [39] and two-layered FNR-tree [21]. In addition, the MON-tree approach [2] further improves the performance of the FNR-tree by representing each edge by multiple line segments (i.e., polyline) instead of just one line segment.

Keeping the current position of moving objects in database is challenging because they change their location frequently. Lots of index structures have been proposed so far to support queries like kNN based on current moving object location. In the LUR-tree [33], the index is not updated if the moving object still lies inside previous minimum bounding rectangle (MBR), i.e., only update the current location. The bottom-up approach [35] is an extension of LUR-tree based on R-trees. LUGrid [54] exploits a lazy insertion and deletion method to handle frequent location updates, where incoming updates having the same to-be-updated disk pages are stored in memory and flushed into disk in batch.

Another important issue is to efficiently find objects that will satisfy some spatial condition at a future time based on their present motion vectors. Some early studies [1, 32] employ dual transformation techniques that represent the predicted positions as points moving in a two-dimensional (2D) space. Recent works focus more on practical implementation, including the R-trees-based TPR-tree [42] and its variations [41] and the B^+-tree-based B^x-tree [29]. But the update performance of the above index mechanisms is not satisfactory. A novel PMR Quad-tree-based index is proposed in [23], which adopts a trajectory segment shared structure while depicting an efficient update algorithm. A dynamic data structure, called adaptive unit, is introduced in [7], which groups neighboring objects with similar movement patterns and captures the movement bounds of the objects based on traffic behavior to reduce updates. A spatial index for the road network is then built over the adaptive unit structures, which forms the ANR-tree [8]. The ANR-tree supports efficient predictive queries and is robust for frequent updates.

1.3.4   Uncertainty Management

As moving objects update their location to server periodically, the server cannot return the exact position of a moving object between two updates, and the only way is to infer the possible position according to the saved trajectory. Such inherent uncertainty has various implications for database modeling, querying, and indexing. Therefore, uncertain management is a very important research issue for moving object databases.

A lot of research has been focused on uncertainty management problem in recent years with many effective models and algorithms being proposed [38, 47, 48]. But most of them are based on Euclidean space in the form of $X \times Y \times T$, while the topology of road network can be used to guide the uncertainty management. Recently, researchers have realized the importance of their interaction, and the modeling of moving objects uncertainty with network constraint has also been studied in [3, 6, 16]. However, a more important problem is how to effectively index the moving object trajectories with uncertainty considered.

1.3.5   Moving Objects Database Querying

Based on moving object data model and indexes, we process the MOD queries to find results. As moving objects have spatial and temporal attributes, spatial and temporal predicates must be indicated to answer queries for moving objects. Hence, there are many types of MOD queries according to different kinds of spatial and temporal predicates.

From the spatial predicates perspective, the typical queries for moving objects mainly include: point query, range query, k-nearest neighbor (kNN) query [31], etc. While from the temporal predicates perspective, queries can be classified into three classes: historical, current, and future queries. Generally, historical queries are usually based on moving object trajectories (e.g., to find moving objects that passed crime region yesterday morning), while current queries and kNN queries are often point query (e.g., to find the vacant taxi closest to a user). Also, all these queries can be divided into Euclidean space-based queries and spatial network-based queries, where different measures of distance are used.

Given that the use of location sensors such as GPS becomes popular, some advanced queries for moving objects database have been increasingly useful. A typical example is similar trajectory query [40], which attempts to find the moving patterns embedded in trajectories, and it has various applications: we can find suspicious objects with abnormal behavior characteristics or discover the movement habit of opponent key sport players through such queries; another example is density query [26, 30] for moving objects in spatial networks, and it provides important real-time traffic information to drivers like the areas with high concentration of moving objects. Also, we further need to monitor the density change and process continuous queries due to the high dynamics of moving objects. In comparison, advanced queries are much more computational intensive than basic queries.

Traditional querying techniques for MOD are based on single database node. However, moving objects usually produce large-scale dynamic data, and we thus have to manage them in database cluster sometimes. As a result, it is necessary to find high-performance querying approaches for processing the above (basic and advanced) queries in distributed moving object databases effectively.

1.3.6 Statistical Analysis and Data Mining of Moving Object Trajectories

Statistical analysis and data mining of moving object trajectories have become important techniques to improve transportation system performance nowadays. By process the traffic data using statistical and mining approaches, knowledge, and intrinsic patterns of the road networks can be discovered and then further used to guide route planning as well as traffic control decision making.

Most of current research on trajectory statistical analysis is based on float car architecture (FCA): Moving objects report its position, speed, and direction to data center, and such data are processed in data center to derive real-time traffic information. The relationship between moving objects data update frequency and statistical results has been discussed, and lots of methods and algorithms have been proposed to address issues like road map matching and real-time traffic parameter computation. However, the FCA-based statistical approaches have some significant drawbacks: (1) The cost of communication and statistical computation based on periodical sampling is very expensive; (2) statistics on discrete float car data impacts the precision and efficiency of traffic flow analysis because it misses the valuable information provided by network topology; and (3) real-time traffic report cannot be supported by FCA-based approaches because of their off-line processing nature. To handle the above problems, Ding and Güting in [15] propose the SBDTN model in which each dynamic moving object data is associated with the underlying road segment and an incremental traffic flow analysis approach using trajectory-based statistical algorithms in [17] based on SBDTN model.

For some new applications, trajectory-based data mining such as real-time clustering analysis is becoming one of the most important requirements, especially, clustering objects in spatial networks [5, 9]. One of the objectives for clustering objects is to identify traffic congestions. A unified framework for clustering moving objects in spatial networks (CMON) is proposed in [5]. The goals are to optimize the cost of clustering moving objects and support multiple types of clusters in a single application. The framework is composed of two components: (1) the continuous maintenance of cluster blocks (CBs) and (2) the periodical construction of clusters with different criteria based on CBs. The network features are explored to reduce the search space and avoid unnecessary computation of network distance.

1.3.7   Location Privacy

Protection of user's privacy has been a central issue for location-based services. Privacy threats related to location-based services are classified into two categories: communication privacy threats and location privacy threats. Location privacy is a particular type of information privacy [4]. In [53], two kinds of privacy protection requirements in LBS are identified: location anonymity and identifier anonymity. To strike a balance between the location privacy and quality of service (QoS), a quality-aware anonymity model for protecting location privacy while meeting user-specified QoS requirements is necessary.

1.4   Applications of Mobile Data Management

The combination of computing techniques and wireless networks makes mobile computing more and more pervasive. Compared with traditional distributed computing environment based on stable networks, mobile computing has the following features: mobility, frequent disconnection, variety of bandwidth, asymmetry of network communication, scalability, limited power of mobile devices, low reliability of the networks, and so on [20]. In such environments, some new technologies in particular the positioning technologies have merged and enabled a variety of new applications such as location-based services.

In mobile computing environments, many new applications deal with a significant amount of data, which leads to the need for mobile data management techniques [18, 27]. Mobile data management mainly includes mobile database techniques, small footprint databases design and implementation, and moving object data management. Mobile database techniques include mobile transaction management, data caching and replication, synchronization, and publication. Small footprint databases techniques include flash-based storage and index model design, query processing and optimization in limited memory, transaction management, recovery techniques, and synchronization. Moving object data management includes modeling and tracking of dynamic location information, uncertainty management, indexing and location-dependent query processing, data mining (e.g., traffic and location prediction), privacy and security, and location dissemination.

In addition, the strong growth in wireless communications and the ever-increasing availability of mobile multipurpose devices have created a global computing environment. People communicate, work, and confer using a wide range of devices all connected via an array of communication networks that provide voice and data access regardless of geographic position. This infrastructure aggregation presents a number of challenges especially when it comes to data-intensive applications such as LBS and PIM and those with sensor networks. Therefore, nontraditional issues including semantics of data, location-centric data services, broadcast and multi-cast delivery, data availability techniques, security of data, as well as privacy questions should be given considerable attention [18, 20].

1.5   Purpose of This Book

Moving object data management is a technically challenging research area that can find various interesting applications in our life. This book gives a comprehensive and complete view of a moving object management system, with the leading techniques of this area to be introduced and discussed. Specifically, it focuses on the hot topics related to MOD, including moving objects modeling, location updating and indexing, querying and prediction for moving objects, uncertainty management, statistical analysis and data mining on moving object trajectories, location privacy, as well as advanced applications in intelligent transportation management and location-based services. It is predictable that moving objects management will become one of the most influential information techniques in the future.

References

1. Agarwal PK, Arge L, Erickson J (2000) Indexing moving points. In: Proceedings of the 19th ACM SIGMOD-SIGACT-SIGART symposium on principles of database systems (PODS 2000), Dallas, pp 175–186
2. Almeida VT, Güting RH (2004) Indexing the trajectories of moving objects in networks. GeoInformatica 9(1):33–60
3. Almeida VT, Güting RH (2005) Supporting uncertainty in moving objects in network databases. In: Proceedings of the 13th annual ACM international workshop on geographic information systems (GIS 2005), Bremen, pp 31–40
4. Beresford AR, Stajano F (2003) Location privacy in pervasive computing. IEEE Pervasive Comput 2(1):46–55
5. Chen J, Lai L, Meng X, Xu J, Hu H (2007) Clustering moving objects in spatial networks. In: Proceedings of the 12th international conference on database systems for advanced applications (DASFAA 2007), Bangkok, pp 611–623
6. Chen J, Meng X (2007) Indexing future trajectories of moving objects in a constrained network. J Comput Sci Technol 22(2):245–251
7. Chen J, Meng X (2009) Update-efficient indexing of moving objects in road networks. GeoInformatica 13(4):397–424
8. Chen J, Meng X, Guo Y, Grumbach S (2007) Indexing future trajectories of moving objects in a constrained network. J Comput Sci Technol 22(2):245–251
9. Chen J, Meng X, Lai C (2007) Clustering objects in road networks (in Chinese). J Softw 18(2):332–344
10. Chen J, Meng X, Li B, Lai C (2006) Tracking network-constrained moving objects with group updates. In: Proceedings of WAIM, Hong Kong, pp 158–169
11. Chon HD, Agrawal D, Abbadi AE (2001) Using space-time grid for efficient management of moving objects. In: Proceedings of the 2nd ACM international workshop on data engineering for wireless and mobile access (MobiDE 2001), Santa Barbara, pp 59–65
12. Civilis A, Jensen CS, Nenortaite J, Pakalnis S (2004) Efficient tracking of moving objects with precision guarantees. In: Proceedings of the 1st annual international conference on mobile and ubiquitous systems, networking and services, Cambridge, pp 164–173
13. Civilis A, Jensen CS, Pakalnis S (2005) Techniques for efficient road-network-based tracking of moving objects. IEEE Trans Knowl Data Eng 17(5):698–712

14. Ding Z, Güting RH (2004) Managing moving objects on dynamic transportation networks. In: Proceedings of the 16th international conference on scientific and statistical database management (SSDBM 2004), Santorini Island, pp 287–296
15. Ding Z, Güting RH (2004) Modeling temporally variable transportation networks. In: Proceedings of DASFAA, Jeju Island, pp 154–168
16. Ding Z, Güting RH (2004) Uncertainty management for network constrained moving objects. In: Proceedings of the 2004 international conference on database and expert systems applications (DEXA 2004), Zaragoza, pp 411–421
17. Ding Z, Huang G (2009) Real-time traffic flow statistical analysis based on network-constrained moving object trajectories. In: Proceedings of DEXA, Linz, pp 173–183
18. Dunham MH, Helal A (1995) Mobile computing and databases: anything new? SIGMOD Rec 24:5–9
19. Forlizzi L, Güting RH, Nardelli E, Schineider M (2000) A data model and data structures for moving objects databases. In: Proceedings of the ACM SIGMOD international conference on management of data, Dallas, pp 319–330
20. Forman GH, Zahorjan J (1994) The challenges of mobile computing. Computer 27:387–403
21. Frentzos E (2003) Indexing objects moving on fixed networks. In: Proceedings of the 8th international symposium on spatial and temporal databases (SSTD 2003), Santorini Island, pp 289–305
22. Gowrisankar H, Nittel S (2002) Reducing uncertainty in location prediction of moving objects in road networks. In: Proceedings of the international conference on information networking, Cheju Island, pp 81–90
23. Guttman A (1984) A dynamic index structure for spatial searching. In: Proceedings of the ACM SIGMOD international conference on management of data (SIGMOD 1984), Boston, pp 47–57
24. Güting RH, Almeida VT, Ding Z (2006) Modeling and querying moving objects in networks. J Very Large Data Bases 15(2):165–190
25. Güting RH, Böhlen MH, Erwig M, Jensen CS, Lorentzos NA, Schneider M, Vazirgiannis M (2000) A foundation for representing and querying moving objects. ACM Trans Database Syst 25(1):1–42
26. Hadjieleftheriou M, Kollios G, Gunopulos D, Tsotras VJ (2003) On-line discovery of dense areas in spatio-temporal databases. In: Proceedings of the 8th international symposium on advances in spatial and temporal databases (SSTD 2003), Santorini Island, pp 306–324
27. Imielinski T, Badrinath BR (1993) Data management for mobile computing. SIGMOD Rec 22:349
28. Jensen CS, Friis-Christensen A, Pedersen TB, Pfoser D, Saltenis S, Tryfona N (2001) Location-based services – a database perspective. In: Proceedings of the 8th Scandinavian research conference on geographical information science (ScanGIS 2001), Ås, pp 59–68
29. Jensen CS, Lin D, Ooi BC (2004) Query and update efficient B^+ tree based indexing of moving objects. In: Proceedings of the 30th international conference on very large data bases (VLDB 2004), Toronto, pp 768–779
30. Jensen CS, Lin D, Ooi BC, Zhang R (2006) Effective density queries on continuously moving objects. In: Proceedings of the 22nd international conference on data engineering (ICDE 2006), Atlanta, p 71
31. Kolahdouzan M, Shahabi C (2004) Voronoi-based K nearest neighbor search for spatial network databases. In: Proceedings of the 30th international conference on very large data bases (VLDB 2004), Toronto, pp 840–851
32. Kollios G, Gunopulos D, Tsotras VJ (1999) Effective density queries on continuously moving objects. In: Proceedings of the 22nd international conference on data engineering (ICDE 1999), Atlanta, p 71
33. Kwon D, Lee SL, Lee S (2002) Indexing the current positions of moving objects using the lazy update R-tree. In: Proceedings of the 3rd international conference on mobile data management (MDM 2003), Singapore, pp 113–120

34. Lam K, Ulusoy Ö, Lee T, Chan E, Li G (2001) An efficient method for generating location updates for processing of location-dependent continuous queries. In: Proceedings of the 7th international conference on database systems for advanced applications, Hong Kong, pp 218–225

35. Lee ML, Hsu W, Jensen CS, Cui B, Teo KL (2003) Supporting frequent updates in R-trees: a bottom-up approach. In: Proceedings of 29th international conference on very large data bases (VLDB 2003), Berlin, pp 608–619

36. Nascimento MA, Silva JRO (1998) Towards historical R-trees. In: ACM symposium on applied computing (SAC 1998), Atlanta, pp 235–240

37. Papadias D, Zhang J, Mamoulis N, Tao Y (2003) Query processing in spatial network databases. In: Proceedings of the 29th international converence on very large data bases (VLDB), Berlin, pp 802–813

38. Pfoser D, Jensen CS (1999) Capturing the uncertainty of moving object representations. In: Proceedings of the 6th international symposium on advances in spatial databases (SSD 1999), Hong Kong, pp 111–132

39. Pfoser D, Jensen CS (2003) Indexing of network constrained moving objects. In: Proceedings of the 11th ACM international symposium on advances in geographic information systems (GIS 2003), New Orleans, pp 25–32

40. Pfoser D, Jensen CS, Theodoridis Y (2000) Novel approaches in query processing for moving object trajectories. In: Proceedings of the 26th international conference on very large data bases (VLDB 2000), Cairo, pp 395–406

41. Saltenis S, Jensen CS (2002) Indexing of moving objects for location-based service. In: Proceedings of the 18th international conference on data engineering (ICDE 2002), San Jose, pp 463–472

42. Saltenis S, Jensen CS, Leutenegger ST, Lopez MA (2000) Indexing the positions of continuously moving objects. In: Proceedings of the ACM SIGMOD international conference on management of data (SIGMOD 2000), Dallas, pp 331–342

43. Sistla P, Wolfson O, Chamberlain S, Dao S (1997) Modeling and querying moving objects. In: Proceedings of the 13th international conference on data engineering (ICDE 1997), Birmingham, pp 422–432

44. Speicys L, Jensen CS, Kligys A (2003) Computational data modeling for network-constrained moving objects. In: Proceedings of the 7th ACM international symposium on advances in geographic information systems, New Orleans, pp 118–125

45. Su J, Xu H, Ibarra O (2001) Moving objects: logical relationships and queries. In: Proceedings of the 7th international symposium on spatial and temporal databases (SSTD 2001), Redondo Beach, pp 3–19

46. Tao Y, Faloutsos C, Papadias D, Liu B (2004) Prediction and indexing of moving objects with unknown motion patterns. In: Proceedings of the ACM SIGMOD international conference on management of data (SIGMOD 2004), Paris, pp 611–622

47. Trajcevski G, Wolfson O, Cao H, Lin H, Zhang F, Rishe N (2002) Managing uncertain trajectories of moving objects with domino. In: Proceedings of the 4th international conference on enterprise information systems (ICEIS 2002), Ciudad Real, pp 769–771

48. Trajcevski G, Wolfson O, Chamberlain S, Zhang F (2002) The geometry of uncertainty in moving objects databases. In: Proceedings of the 8th international conference on extending database technology: advances in database technology (EDBT 2002), Prague, pp 233–250

49. Vazirgiannis M, Wolfson O (2001) A spatialtemporal model and language for moving objects on road networks. In: Proceedings of the 7th international symposium on spatial and temporal databases, Redondo Beach, pp 20–35

50. Wolfson O, Sistla A, Chamberlain S, Yesha Y (1999) Updating and querying databases that track mobile units. Distrib Parallel Databases 7(3):257–387

51. Wolfson O, Xu B, Chamberlain S, Jiang L (1998) Moving object databases: issues and solutions. In: Proceedings of the 10th international conference on scientific and statistical database management (SSDBM 1998), Capri, pp 111–122

52. Wolfson O, Yin H (2003) Accuracy and resource consumption in tracking and location prediction. In: Proceedings of the 7th international symposium on spatial and temporal databases (SSTD 2003), Santorini Island, pp 325–343
53. Xiao Z, Meng X, Xu J (2007) Quality aware privacy protection for location-based services. In: Proceedings of the 12th international conference on database systems for advanced applications (DASFAA 2007), Bangkok, pp 434–446
54. Xiong X, Mokbel MF, Aref WG (2006). LUGrid: update-tolerant grid-based indexing for moving objects. In: Proceedings of the 7th international conference on mobile data management (MDM 2006), Nara, p 13

Chapter 2
Moving Objects Modeling

Abstract Location modeling is the foundation for moving objects databases. Existing database management systems are not well equipped to handle continuously changing data, such as the position of moving objects. The reason for this is that in traditional databases, data is assumed to be constant unless it is explicitly modified. This is unsatisfactory for MOD since locations of moving objects are continuously changing. In this chapter, we overview some representative models for moving objects and present two moving object models that are based on the concept of dynamic transportation networks.

Keywords Location modeling • Dynamic transportation network moving objects model • Moving object trajectory • Spatial network • Moving object database

2.1 Introduction

The continuous advances in wireless sensor networks and position technologies enable traffic management and location-based services that track continuously changing positions of moving objects. Timely location information is becoming one of the key features of these applications. In existing DBMSs, data is assumed to be constant unless it is explicitly modified. Therefore, the continuously changing data, such as the location of moving objects, are hard to handle.

In order to represent trajectories of moving objects in databases and to answer queries about their position, a straightforward way is to update the position of moving objects continuously. This is not feasible yet because of the excessive I/O and wireless-bandwidth cost if updates are frequent enough to guarantee high tracking precision. Also, due to the disconnections, it is not possible for an object to continuously update its position to the server. Therefore, new location modeling methods are needed to solve this problem.

Figure 2.1 illustrates this dilemma. As Fig. 2.1a shows, when the moving object frequently reports its location updates to the database server, the trajectories are

X. Meng et al., *Moving Objects Management: Models, Techniques and Applications*, DOI 10.1007/978-3-642-38276-5_2,
© Tsinghua University Press, Beijing and Springer-Verlag Berlin Heidelberg 2014

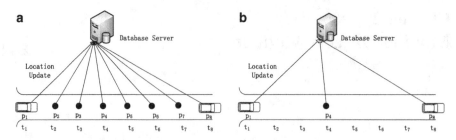

Fig. 2.1 Challenge of moving objects modeling. (**a**) Query (t_3) = p_3. (**b**) Query (t_3) =?

stored as intensive samples in database and queries like: what is the location of the moving object at t_3 can be preciously answered, i.e., p_3. It however has such a high update cost that the database server cannot bear simultaneous updates sent from a large number of moving objects. On the other hand, as Fig. 2.1b, when the moving objects make location updates for a long time interval, the trajectories are stored as sparse samples and the same queries cannot be preciously answered at the most time, though the workload of database server is reduced. Such dilemma is caused by the conflict between the continuous movement of objects and the discrete storage in databases. Therefore, it is crucial to find a proper balance between updating cost and tracking precision.

So far, many models and algorithms have been proposed to handle the continuously changing positions of moving objects. Basically, each attribute of a moving object can be modeled either statically or dynamically. A static attribute only changes when an explicit update of the database occurs; in contrast, a dynamic attribute changes over time according to some functions, even if it is not explicitly updated. For example, assume that a moving object whose position at any point in time is given by a (x, y) coordinate sequence in two-dimensional space, each of the object's coordinates is a dynamic attribute. The main difference between the dynamic attribute of moving objects and attributes in traditional database systems is that the values of (dynamic) location attribute are continuously changing.

Wolfson et al. in [15, 20] first propose a moving objects spatio-temporal (MOST) model, which represents the location as a dynamic attribute. Later, models based on linear constraints [17], abstract data types [9], and space-time grid storage [3] for moving objects are proposed. In these modeling approaches, attributes of the moving objects are dynamic and change continuously as a function of time, avoiding explicitly updates. Corresponding data types and operations are defined with considerations of dynamic attributes to provide a feasible DBMS model capable of handling such time-dependent characteristic of moving object. To retrieve and query the moving objects, various sorts of functions are proposed, such as: direct interpolation or exploration, linear function based, and nonlinear function based. However, in most real-life applications, objects move within constrained networks, especially in the case of transportation networks (e.g., vehicles move on road networks). These models do not take into account the interaction between moving objects and the underlying transportation networks.

The interaction is another very important aspect in the management of moving objects. For example, in location tracking, the road-network representation of moving objects can be exploited to reduce the number of updates from moving objects to the database server [4]. For indexing moving objects in road networks, the temporal aspect can be distinguished and related to the road network to save considerable index storage space [1, 7], since the spatial property of objects' movement is already captured by the network. In addition, by using the network constraints, the query processing can also be improved [10, 14]. Hence, a lot of models connecting moving objects with the road-network representation have been proposed [8, 13, 16, 19]. Considering the underlying road network has brought a lot of benefits in indexing and querying. However, it also leads to the difficulties to model the road network and the moving object simultaneously. The major reason is that the road network should be viewed dynamically also. In a road network, a route can have different states, like: blocked, opened, or closed, etc. Moreover, different events can happen on the road network, such as: traffic jams, car accidents, road construction, or road repairs, etc. Therefore, it is necessary to dynamically model the underlying road network. Based on that consideration, Ding and Güting [5] have proposed a method to model the moving object on dynamic road network. In the scenario of a realistic traffic system, especially in intelligent transportation system which is popular nowadays, the model needs to reflect real-time traffic conditions of the road networks. The goodness of reflecting real-time traffic conditions includes two folds. For one thing, precise route planning requires full facilitation of the GPS samples. For another thing, dynamic is an inherent characteristic of moving object and road network which both require corresponding dynamic model.

More recently, the graph of cellular automata (GCA) is proposed in [2] to integrate the traffic movement features into the model of moving objects and the underlying road network. The GCA model exploits the stochastic behavior of the real traffic by the cellular automaton which is used in the traffic simulation. It also combines the road-network model with the real movement of objects and therefore can support the management of network-constrained moving objects.

In this chapter, we first give a brief introduction of some representative models of moving objects and then present a moving object model called DTNMOM which considers the dynamics of underlying road network. Finally, a DTNMOM based on atomic route section model called ARS-DTNMOM is further presented, which introduces the concept of atomic route section and defines its corresponding data types and operations in database. One noteworthy feature shared by the two models is that both of them can reflect real-time traffic conditions of the road network.

2.2 Representative Models

2.2.1 Moving Object Spatio-Temporal (MOST) Model

Wolfson et al. in [15, 20] first propose a moving objects spatio-temporal (MOST) model for databases with dynamic attributes, i.e., attributes that change continuously as a function of time, without being explicitly updated. In the MOST model, the location as a dynamic attribute is represented as a function of time. Formally, a dynamic attribute A is represented by three sub-attributes, $A.value$, $A.updatetime$, and $A.function$, where $A.function$ is a function of a single variable t that has value 0 at $t = 0$. The value of a dynamic attribute depends on the time, and it is defined as follows. At time $A.updatetime$, the value of A is $A.value$, and until the next update of A, the value of A at time $A.time + t_0$ is given by $A.value + A.function(t_0)$. An explicit update of a dynamic attribute may change its value sub-attribute, or its function sub-attribute, or both sub-attributes. For example, the position of a car is given as a function of its motion vector (e.g., north, at 60 miles/h). In other words, it considers a higher level of data abstraction, where an object's motion vector (rather than its position) is represented as an attribute of the object. Obviously, the motion vector of an object can change (and thus it can be updated), but in most cases it does so less frequently than the position of the object. When a dynamic attribute is queried, the answer returned by the MOD consists of the value of the attribute at the time the query is entered. In this sense, the MOST model is different from existing database systems, since, unless an attribute has been explicitly updated, a DBMS returns the same value for the attribute, independent of the time at which the query is posed. With the motion vector, the MOST model is capable of representing not only the current but also the near-future position of moving objects.

However, due to the limited expression ability of the simple function in dynamic attributes, the MOST model can only represent the future positions of moving objects in a short period. The study [6] solves this problem by presenting the moving object's discrete data model, in which the complicated trajectory of a moving object can be represented by a set of relatively simple discrete segments. In addition, Su et al. in [17] present a data model for moving objects based on linear constraint databases. Chon et al. in [3] propose a space-time grid storage model for moving objects. In [9], Güting et al. present a data model and data structures for moving objects based on abstract data types. These studies focus on the modeling of objects moving in free spaces, not constrained by any spatial network.

2.2.2 Abstract Data Type (ADT) with Network

Take the abstract data type (ADT) proposed by Güting et al. as an example. The goal of abstract data type is to provide a DBMS data model and query language capable of handling such time-dependent geometries, including those changing continuously that describe moving objects. Two fundamental abstractions in ADT are moving point and moving region. Moving point describes the time-dependent position of a moving object, and moving region describes the time-dependent extent of a moving object. ADT represents such time-dependent geometries as attribute data types with suitable operations to provide an abstract data type extension to a DBMS data model and query language. Besides the main types of ADT, moving point and moving region, a relatively large number of auxiliary data types are included. For example,

ADT includes a line type to represent the projection of a moving point into the plane or a "moving real" to represent the time-dependent distance of two moving points. ADT has three advantages: (1) achieves orthogonally in the design of the type system, i.e., type constructors can be applied uniformly; (2) genericity and consistency of operations, i.e., operations range over as many types as possible and behave consistently; and (3) closure and consistency between structure and operations of nontemporal and related temporal types. Satisfying these goals leads to a simple and expressive system of abstract data types that can be integrated into a query language to yield a powerful language for querying spatio-temporal data, including moving objects.

For moving objects in a spatial network, when adding the network constraint, we need to consider not only the location representation but also the modeling of the spatial network as well as the spatial objects. To represent moving objects in road network, ADT adopts the graph representation which gives the definitions of routes and junctions and offers two kinds of routes called simple and dual routes. There are also two concepts for positions on roads called route measure and route location. The route measure is independent from the kind of route (simple or dual); it is just a distance from the origin of the route. Junctions between two routes are positioned at two distinct route measures. The route location depends on the route type. For a simple route, it is the same as the route measure; for a dual route, it is a route measure plus a flag from the set $\{up,down\}$. Similarly, on a simple route, a route interval is given by two measures, on a dual route by two measures plus an up-down flag. Hence, a route description consists of an identifier, a length, a curve describing its geometry in the plane, a route type, and a flag indicating how route locations are to be embedded into space. The geometry is a simple, non-self-intersecting curve in the plane which may be open or closed (a cycle). It is represented by a polyline. In spatial databases and also in ADT framework, a data type _line_ is available that can represent such values.

The presented abstract data types can now be embedded into any DBMS data model as attribute data types, and the operations be used in queries.

Let us use a junction in road network for illustration. A junction in road network is a triple consisting of two route measures in road network with distinct route identifiers and a connectivity code, an integer value encoding which movements through the junction are possible. At a junction between routes A and B, various transitions may be possible or not for a moving object, for example, a transition $A_{up} \rightarrow B_{up}$. This is illustrated in Fig. 2.2a, b. In most cases, junctions are built to allow for all eight possible transitions from one route to the other. However, this is not always true. Figure 2.2a shows an example of a physical highway junction where in fact only the transitions $A_{up} \rightarrow B_{up}$, $A_{up} \rightarrow B_{down}$, and $A_{down} \rightarrow B_{down}$ are possible. ADT can represent the possible transitions in a 4×4 matrix as shown in Fig. 2.2c. The 1's in the diagonal represent the fact that it is possible to stay on a route in the same direction. A transition such as, for example, $A_{up} \rightarrow A_{down}$ would be set to 1 if a U-turn were possible. In general, for the definition of the transition matrix, A and B are chosen such that the route identifier of A is smaller than that of B.

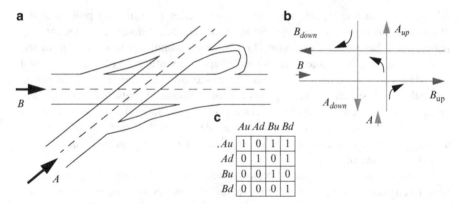

Fig. 2.2 (**a**) A physical highway junction, (**b**) its diagrammatical representation, (**c**) the transition matrix

2.2.3 Graph of Cellular Automata (GCA)

The graph of cellular automata (GCA) [2] integrates the traffic movement features into the model of moving objects and the underlying road network, which models the road intersections as graph nodes and the road segments with no intersections as graph edges. Different from the general graph model, each edge in GCA consists of a cellular automaton (CA), which is represented, in a discrete mode, as a finite sequence of cells. Each cell corresponds in practice to some road segment of about 7.5 m. Cellular automata were originally introduced by von Neumann [12] and Ulam [18] in the 1960s with the particular purpose of modeling biological self-reproduction. Since then, they have been used broadly for physics applications such as particle transport simulations and thermodynamics studies. The CA model was used in this context in [11] because in the CA model it is quite simple and easy to describe the interaction of cells; it is suitable for computer simulations of discrete phenomena. A moving object is represented as a symbol attached to the cell in the GCA, and it can move several cells ahead at each time unit. The motion of an object is represented as information in the form (time, location). Representing such information of a moving object as a trajectory is a typical approach [19]. In the GCA model [2], the trajectory of a moving object can be divided into two types: the in-edge trajectory for the object's movement in one edge (CA) and the global trajectory for the object that may move cross several edges (CAs) during its movement.

Figure 2.3 shows an example of a road network and its GCA model. Each node has a label that represents an intersection of the road network. The wide lines represent edges and each edge treated as one CA connects many cells.

A graph of cellular automata can be formally defined as follows.

Definition 2.1. The structure of a GCA is a directed weighted graph $G = (V, E, l)$ where V is a set of vertices (i.e., nodes), E is a set of edges, and $l : E \to \mathbb{N}$ is a function that associates to each edge the number of cells of the corresponding cellular automaton.

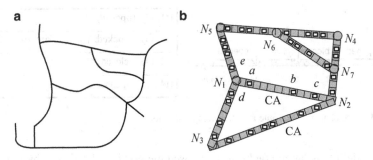

Fig. 2.3 An example of a road network and its GCA model. (**a**) A road network. (**b**) An instance of GCA

We assume a countably infinite alphabet Ω : $\{\alpha, \beta, \gamma, \cdots\}$, denoting moving objects' names. Let C be the set of cells of a GCA.

A configuration or an instance of a GCA is a mapping from the cells of the GCA to constants in Ω together with a given velocity. Intuitively, the velocity is the number of cells an object can traverse during a time unit.

Definition 2.2. An instance I of a GCA is defined by two functions:

$\mu : C \rightarrow \Omega \bigcup \{\emptyset\}$ (1−1 mapping).

$v : \Omega \rightarrow \mathbb{N}$.

A moving object is represented as a symbol attached to the cell in the GCA, and it can move several cells ahead at each time unit. Figure 2.3b is an instance of the GCA corresponding to the road network of Fig. 2.3a. In Fig. 2.3b, moving objects are denoted by squares. A moving object lies on exactly one cell of the edge and its location can be obtained by computing the number of cells relative to the start node. For instance, the object α lies on the edge (N_1, N_2) and it is two cells away from N_1 along the edge. Therefore, its position can be expressed by $(N_1, N_2, 2)$.

2.3 DTNMOM

The route-based dynamic transportation networks moving objects model (DTN-MOM) [5] is composed of two steps. The first step is to model the underlying dynamic transportation networks with traffic state, and the second one is the modeling of moving objects on transportation network. In DTNMOM, the transportation network is modeled as "dynamic" graphs, so that we can express traffic state changes (e.g., from unobstructed state to congested state) and topology changes (e.g., insertion and deletion of junctions or routes) easily. For simplicity, "dynamic transportation networks" and "dynamic graphs" will be used interchangeably throughout this chapter.

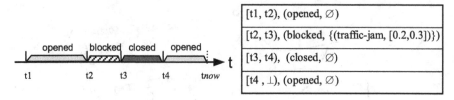

Fig. 2.4 State changes of a route and their temporal units

In modeling dynamic graphs, a state-based method is utilized. The basic idea is to associate a temporal attribute to every route or junction of the graph system, so that the state of the route or junction at any time instant can be retrieved. Since the changes to the graph system are discrete, we can use a series of temporal units to represent a temporal attribute with each temporal unit describing one single state of the route or junction during a certain time period. In this way, the whole spatio-temporal history of the graph system can be stored and queried.

As for the modeling of movement of objects, since in most cases a moving object can be viewed as a point, they are modeled as moving graph points in the dynamic transportation networks. A moving graph point is a function from time to graph point, which can be represented as a group of moving units in the discrete model. The methodology proposed in DTNMOM can be easily extended to deal with more complicated situations where moving objects need to be modeled as moving graph lines or moving graph regions.

Definition 2.3. The temporal attribute associated with a junction or a route, denoted by tp, describes the state history of the junction or route, which is defined as a sequence of the following form:

$$tp = ((I_i, s_i))_{i=1}^n$$

where I_i is a time interval and s_i is the state of the junction or route during I_i. $(I_i, s_i)(1 \leq i \leq n)$ is called the ith temporal unit of tp.

For a certain temporal unit (I_i, s_i) $(1 \leq i \leq n)$, I_i is composed of two time instant values, $min(I_i)$ and $max(I_i)$, which indicate the starting point and the end point of I_i, respectively. $min(I_i)$ must be a defined value, while $max(I_i)$ can be either defined or undefined. If $max(I_i)$ is an "undefined" value $\perp$, then I_i is called an open temporal unit. Otherwise, it is called a closed temporal unit. Semantically, $\perp$ means "until now." Therefore, if a junction or route is still active in the transportation network, its temporal attribute will contain exactly one open temporal unit, which forms its last temporal unit. Otherwise, if it has already been deleted from the transportation network, then its temporal attribute will only contain closed temporal units.

The insertion and deletion time of a junction or a route can be decided by $min(I_1)$ and $max(I_n)$, respectively. Figure 2.4 illustrates an example temporal attribute value.

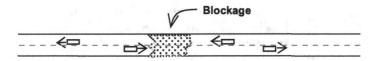

Fig. 2.5 A blocked route with moving objects

Definition 2.4. A state of a junction or a route, denoted by s, is defined as follows:

$$s = (\delta, (br_i, BP_i)_{i=1}^{n})$$

where $\delta \in \{opened, closed, blocked\}$. If $\delta = blocked$, then s must be associated with a route, and $(br_i, BP_i)_{i=1}^{n}$ is needed in this situation to describe the blockages of the route where br_i describes the reason (traffic jam, construction, traffic control, etc.) and $BP_i \subseteq [0, 1]$ describes the location of the ith blockage of the route.

In the above definition, we assume that the location of the blockage is static so that it can be expressed as a closed interval over $[0, 1]$, whose boundaries indicate the location of the borders of the blockage.

In dynamic transportation networks, there are two possible states of each junction, opened and closed, and a route has three possible states: opened, closed, and blocked. If a junction or a route is opened, then it is entirely available to moving objects. If a junction or a route is closed, then it is entirely unavailable to moving objects, which means that no moving objects are allowed to stay or move on any part of it. A closed junction or route is not deleted from the system. Instead, it is only temporarily unavailable to moving objects and can be reopened afterwards.

The blocked state is used to describe a special kind of state of a route, which means "partially available" to moving objects. That is, the unblocked part of the route is still available to moving objects, but no moving objects can move through the blocked part. Figure 2.5 gives an example of blocked route.

In the dynamic graph system, since every junction or route has a temporal attribute associated, we can know its state at any given time instant. This is very useful in moving objects databases since a lot of queries can only be processed efficiently by accessing the states of the transportation networks. For instance, "please tell me all the routes which are currently blocked by traffic jams and the moving objects affected by them." Besides, through the temporal attribute, we can also know the life span of any junction or route of the graph system so that the topology changes of the transportation networks can also be expressed and queried. For instance, "find the shortest path from a to b at time instant t."

Definition 2.5. A dynamic route, denoted by r, is defined as follows:

$$r = (gid, rid, route, len, fdr, tp)$$

where gid and rid are identifiers of the road networks and the route, respectively, *route* is a polyline which describes the geometry of r, *len* is the length of the route, $fdr \in \{0, 1, 2\}$ is the traffic flow directions allowed in the route, and tp is the temporal attribute associated with r.

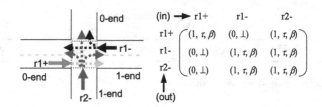

Fig. 2.6 A junction and its connectivity matrix

The polyline route in the above definition can be defined as a series of points in the Euclidean space. For simplicity, we suppose that the graph system is spatially embedded in the $X \times Y$ plane so that the polyline can be presented as a series of points in the $X \times Y$ plane. The polyline is considered directed, whose direction is from the first vertex to the last vertex, which enables us to speak of the beginning point (or 0-end) and the end point (or 1-end) of the route.

The traffic flow directions allowed in a route can have three possibilities that are specified by fdr, whose value can assume 0, 1, 2, which corresponds to "from 0-end to 1-end," "from 1-end to 0-end," and "both directions allowed," respectively.

There will be multiple graphs coexisting, while each graph is composed of a set of routes and a set of junctions. For each route, its geometry is described by a polyline so that it can actually be assumed as an arbitrary shape instead of just a straight line. A junction connects two or more routes of the graph system. The connected routes can come from one graph (the junction is called "in-graph junction") or belong to different graphs (the junction is called "inter-graph junction").

Definition 2.6. A dynamic in-graph junction of graph G, denoted by j, is defined as follows:

$$j = (gid, jid, loc, ((rid_i, pos_i))_{i=1}^n, m, tp)$$

where gid and rid are identifiers of the road networks and the route, respectively, loc is the location of j which can be presented as a point value in the $X \times Y$ plane, $((rid_i, pos_i))_{i=1}^n$ describes the routes connected by j, m is the connectivity matrix of j, and tp is the temporal attribute associated with j.

(rid_i, pos_i) in the above definition indicates the ith route connected by j, where rid_i is the identifier of the route and $pos_i \in [0, 1]$ describes the position of the junction inside the route. We suppose that the total length of any route is 1, and then every location in the route can be presented by a real number $p \in [0, 1]$.

The matrix m describes the connectivity of the junction. It contains possible matches of traffic flows in the routes connected by the junction, and the element value associated with each match can assume either 0 or 1, which indicates whether moving objects can transfer from the "in" traffic flow to the "out" traffic flow through this junction, as shown in Fig. 2.6.

Definition 2.7. A dynamic inter-graph junction, denoted by j^*, is defined as follows:

$$j^* = (jid, loc, ((gid_i, rid_i, pos_i))_{i=1}^n, m, tp)$$

The definition of the inter-graph junction is very similar to that of the in-graph junction. The 3-tuple $(gid_i, rid_i, pos_i)(1 \leq i \leq n)$ describes the routes connected by j^*, which comes from different graphs.

Definition 2.8. A dynamic graph, G, is defined as a pair:

$$G = R, J$$

where R is a set of dynamic routes and J is a set of dynamic in-graph junctions.

Definition 2.9. A dynamic graph system, GS, is defined as a set of dynamic graphs and inter-graph junctions:

$$GS = G_1, G_2, \cdots, G_n, j_1^*, j_2^*, \cdots, j_m^*$$

where $n \geq 1, m \geq 0, G_i(1 \leq i \leq n)$ is a dynamic graph and $j_k^*(1 \leq k \leq m)$ is an inter-graph junction.

Definition 2.10. A moving graph point, mgp, is defined as a function from time to graph point, that is,

$$mgp = f : T \rightarrow GP$$

where T is the domain of time and GP is the domain of graph point of the graph system.

To apply the model to the low-sampling-rate samples, a moving graph point can also be expressed as a set of moving units, and each moving unit describes one single moving pattern of the moving object for a certain period of time.

Definition 2.11. A discrete presentation of moving graph point, $dmgp$, is defined as a sequence:

$$dmgp = ((t_i, (gid_i, rid_i, pos_i), vm_i))_{i=1}^n$$

where t_i is a time instant, $(gid_i, rid_i, pos_i) = gp_i$ is a graph point describing the location of the moving object at time t_i, and vm_i is the speed measure of the moving object at time t_i, and $(t_i, (gid_i, rid_i, pos_i)) = \mu_i$ is called the ith moving unit of $dmgp$.

The speed measure vm is a real number value. Its abstract value equals to the speed of the moving object, while its sign (either positive or negative) depends on the direction of the moving object. If the moving object is moving from 0-end towards 1-end, then the sign is positive. Otherwise, if it is moving from 1-end to 0-end, the sign is negative.

2.4 ARS-DTNMOM

To model the moving object and the underlying dynamic transportation networks, it is crucial to describe the concept of route precisely. For each route, its geometry is described by a polyline so that it can actually assume an arbitrary shape instead of just a straight line. The polyline is considered as directed, whose direction is from the first vertex to the last, which enables us to speak of the beginning point (or 0-end) and the end point (or 1-end) of the route. A route is composed of a set of *directed atomic route sections* (ARS). Conceptually, a directed atomic route section is a directed segment of the traffic flow within a route, which connects two junctions and does not contain any other junctions from which moving objects can exit. The direction of ARS is indicated by the order of the two junctions (from the first junction toward the second). Since ARS is the basic unit for navigation, we choose it as the basic unit for keeping traffic states and parameters. Figure 2.7 illustrates the atomic route sections in the ARS-DTNMOM.

In Fig. 2.7, route $r1$ has two traffic flow directions, "+" and "−" ("+" means that traffic flow is from 0-end to 1-end and "−" is from 1-end to 0-end; see Definition 2.2). Therefore, the directed atomic route sections are also in two directions. Some ARSs are symmetrical (say, $ars1$ and $ars10$), but essentially, ARSs of the two route sides can be asymmetrical (e.g., $ars2$ and $ars9$, $ars3$ and $ars8$).

Additionally, to support the realization of ARS-DTNMOM, the underlying transportation network is modeled as a dynamic graph so that the state, topology, and traffic parameters of the transportation graph at any time instant can be tracked and queried. Moving objects are modeled as moving graph points which move only within predefined transportation networks. The data model is given as a collection of data types and operations which can be plugged as attribute types into a DBMS to obtain a complete data model and query language.

The above ARS-DTNMOM can describe complicated traffic network and the traffic flows in it. The benefit of this network framework is twofold. First, since routes are explicitly presented, the MOD system can take routes as the basic unit for network position representation and for location updates. Therefore, if a moving

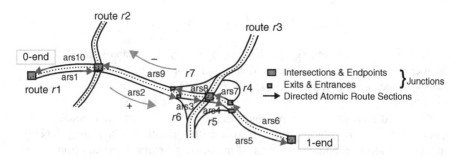

Fig. 2.7 ARS-based DTNMOM

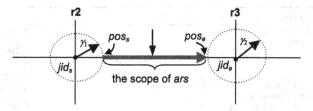

Fig. 2.8 Directed atomic route section

object moves in a certain route with roughly steady speed, no location updates will be triggered even though it may pass through several junctions along the route. As a result, the location update cost can be reduced. In this aspect, it is superior to the DTNMOM. Second, through atomic route sections, we can describe traffic parameters and state information in more detail so that important utilities such as traffic aware navigation can be better supported, and in this aspect, it is superior to the DTNMOM.

Definition 2.12. A directed atomic route section (ARS), denoted by *ars*, is a directed edge which connects two nearby junctions of the network and does not contain any other junctions in between along the same traffic flow direction, which is defined as

$$ars = (aid, (jid_s, pos_s), (jid_e, pos_e))$$

where $aid \in int$ is the identifier of *ars* and (jid_s, pos_s) and (jid_e, pos_e) describe the starting point and the end point of *ars*, respectively, where jid_s and jid_e are junction identifiers and pos_s and pos_e are the starting position and end position of *ars* which are measured just outside the junction area borders, as illustrated in Fig. 2.8.

Figure 2.9 presents the type system of the ARS-DTNMOM. Type constructors listed in Group 1 are basic ones which have been defined and implemented in [6]. In the following, we mainly focus on the type constructors listed in Group 2. Some type constructors in Group 2, such as graph state data types, have been discussed in [5], but most of them are modified to fit into the atomic route section-based moving object model in dynamic transportation network.

Among the type constructors listed in Group 2, graph state (GSTATE) data types, graph temporal (GTEMPORAL) data types, and dynamic graph (GRAPH) data types are used to describe dynamic transportation networks. Graph spatial (GSPATIAL) data types describe static graph objects, which form the basis for the modeling and querying of moving objects. Temporal (TEMPORAL) data types describe moving objects.

As shown in Fig. 2.9, temporal data types are obtained by "extending" the previously defined type constructors moving and intime so that the gpoint data type is included as their arguments. SPATIAL data types are still reserved for these two

Group	Type constructor	Signature	
	int, real, string, bool		→ BASE
	point, points, line, region		→ SPATIAL
1	*instant*		→ TIME
	range	BASE ∪ TIME	→ RANGE
	intime, moving	BASE ∪ SPATIAL	→ TEMPORAL
	status, blockage, blockreason, blockpos, traffpara, traffparas, state		→ GSTATE
	intimeevent, intimestate, g_temporal		→ GTEMPORAL
2	*dynroute, dynjunct, ars*		→ GRAPH
	gpoint, gpoints, grsect, gline, gregion		→ GSPATIAL
extending→	*intime, moving*	{*gpoint*} ∪ BASE ∪ SPATIAL	→ TEMPORAL

Fig. 2.9 Signatures describing the type system of ARS-DTNMOM

type constructors to deal with the situations when moving objects move outside of the predefined transportation network, for instance, moving in a lake or in a big square.

In ARS-DTNMOM, the transportation network is modeled as a dynamic graph, with every junction or route associated with a g_temporal attribute which describes its state history.

Definition 2.13. A polyline can be expressed by a sequence of points which correspond to the vertexes of the polyline. Therefore, we can define polylines as follows:

$$polyline = < p_1, p_2, \cdots, p_n > |n \geq 2, \forall i \in 1, \cdots, n : p_i \in D_{point}$$

Definition 2.14. A dynamic route in ARS-DTNMOM can be viewed as a normal graph route with a temporal attribute associated. The carrier set of the *dynroute* data type is defined as follows:

$$D_{dynroute} = \{(rid, geo, (jid_i, pos_i)_{i=1}^n, len, ARS, tp)\}$$

where *rid* is the identifier of the route which is isomorphic to integer; *geo* is a polyline which describes the geographical shape of the route; $(jid_i, pos_i)(1 \leq i \leq n)$ indicates the ith junction associated with the route, where jid_i is the identifier of the junction and $pos_i \in [0, 1]$ describes its position inside the route (we suppose that the total length of the road section is 1, and then any location in the road section can be represented by a real number $p \in [0, 1]$); *len* is the length of the route; *ARS* is the set of atomic route sections of the route; and $tp \in D_{g\_temporal}$ is the graph temporal attribute associated with the route.

In the earlier work on moving objects databases [6], Güting et al. have defined and implemented a rich set of operations on the data types. In this subsection, we show how these predefined operations are to be systematically adapted to the ARS-DTNMOM by an "Extending" technique. The basic idea of the "Extending" is to add the newly introduced data types into the signatures of the previously defined

Group	Class	Operations
Non-Temporal	Predicates	isempty, =, ≠, <, ≤, >, ≥, intersects, inside, before touches, attached, overlaps, on_border, in_interior
	Set operations	intersection, union, minus, crossings, touch_points, common_border
	Aggregation	min, max, avg, avg[center], single
	Numeric	no_components, size, perimeter, size[duration], size[length], size[area]
	Distance & direction	distance, direction
	Base type specific	and, or, not
Temporal	Projection to Domain/Range	deftime, rangevalues, locations, trajectory, routes, traversed, inst, val
	Interaction with Domain/Range	atinstant, atperiods, initial, final, present, at, atmin, atmax, passes
	When	when
	Lifting	(All new operations inferred)
	Rate of Change	derivative, speed, turn, velocity
Graph Specific	Transformation	graph_euc, euc_graph, getjunct, getroute, getid
	Construction	gpoint, grsect
	Data Extraction	geo, g_temporal, g_atinstant, xstate, xstatus, xblocs, xtraffpara
	Truncation	g_atperiods, g_present
	Projection	g_deftime

Fig. 2.10 Operations of ARS-DTNMOM

operations so that the extension of these operations can be expanded to include the newly defined data types. In addition to the extended operations, we will also define a set of new operations, which are mainly focused on graph-specific data types. Figure 2.10 gives a summary of the operations in the ARS-DTNMOM.

The ARS-DTNMOM operations are extended from [6]. The general rules for extending can be summarized as follows:

1. Every operation whose signature involves point is extended to include gpoint.
2. Every operation whose signature involves line is extended to include grsect and gline.
3. Some of the operations whose signature involves region are extended to include gregion.
4. Operations which are only suited for 1D data types (see [6]) and for some specific data types (such as region) are not extended.

According to the above rules, the underscored (line- and dot-underscored) operations in Fig. 2.10 are extended, while other operations are not affected.

The nontemporal operations are listed in Fig. 2.10. Previously, the signatures of most nontemporal operations (see the line-underscored nontemporal operations, such as isempty) are defined with two data type variables π and δ, where $\pi \in$ *int, bool, string, real, instant, point* and $\delta \in$ *range(int), range(bool), range(string), range(real), periods, points, line, region*. In ARS-DTNMOM, we extend the domain of π and δ like this:

$\pi \in$ *int, bool, string, real, instant, point* ∪ *gpoint, status* $\delta \in$ *range(int), range(bool), range(string), range(real), periods, points, line, region* ∪ *gpoints, grsect, gline, gregion*.

As a result of this change, all operations whose signatures are defined with π, δ variables are extended to cover the newly introduced data types in ARS-DTNMOM and DTNMOM. As for operations whose signatures are not defined with π, δ variables (see the dot-underscored nontemporal operations in Fig. 2.10, such as *crossings*), supplement of their signatures can be made by including the newly introduced data types while keeping their original semantics.

Meanwhile, for the temporal operations listed in Fig. 2.10, the extension can be made in a similar way. The signatures of most temporal operations (see the line-underscored temporal operations, such as *deftime*) are defined by two data type variables α and β, (α, β $\in$ *BASE* $\bigcup$ *SPATIAL*). The domain of α and β can be extended like this: α $\in$ *BASE* $\bigcup$ *SPATIAL* $\bigcup$ *gpoint*, β $\in$ *BASE* $\bigcup$ *SPA-TIAL* $\bigcup$ *GSPATIAL*.

2.5   Summary

For managing moving objects in a spatial network, the challenging first step is to precisely represent and model their locations. In this chapter, we first introduce a few representative models for moving objects databases. Then we present a DTNMOM, where the transportation network is modeled as "dynamic" graphs, so that traffic state changes and topology changes can be easily expressed. Afterwards, the atomic route section-based DTNMOM (ARS-DTNMOM) which describes the concept of route more precisely in atomic route sections is introduced finally.

References

1. Almeida VT, Güting RH (2005) Indexing the trajectories of moving objects in networks. GeoInformatica 9(1):33–60
2. Chen JD, Meng XF, Guo YY, Grumbach S, Sun H (2006) Modeling and predicting future trajectories of moving objects in a constrained network. In: 7th international conference on mobile data management (MDM 2006), Nara, pp 156–166
3. Chon HD, Agrawal D, Abbadi AE (2001) Using space-time grid for efficient management of moving objects. In: Proceedings of the 2nd ACM international workshop on data engineering for wireless and mobile access (MobiDE 2001), Santa Barbara, pp 59–65
4. Civilis A, Jensen CS, Pakalnis S (2005) Techniques for efficient road-network-based tracking of moving objects. IEEE Trans Knowl Data Eng 17(5):698–712
5. Ding Z, Güting RH (2004) Modeling temporally variable transportation networks. In: Proceedings of the 9th international conference on database systems for advanced applications (DASFAA 2004), Berlin, pp 154–168
6. Forlizzi L, Güting RH, Nardelli E, Schneider M (2000) A data model and data structures for moving objects databases. In: Proceedings of the ACM SIGMOD international conference on management of data (SIGMOD 2000), Dallas, pp 319–330
7. Frentzos E (2003) Indexing objects moving on fixed networks. In: Proceedings of the 8th international symposium on spatial and temporal databases (SSTD 2003), Santorini Island, pp 289–305

8. Güting RH, Almeida VT, Ding Z (2006) Modeling and querying moving objects in networks. VLDB J 15(2):165–190

9. Güting RH, Böhlen MH, Erwig M, Jensen CS, Lorentzos NA, Schneider M, Vazirgiannis M (2000) A foundation for representing and querying moving objects. ACM Trans Database Syst 25(1):1–42

10. Kolahdouzan M, Shahabi C (2004) Voronoi-based K nearest neighbor search for spatial network databases. In: Proceedings of the 30th international conference on very large data bases (VLDB 2004), Toronto, pp 840–851

11. Nagel K, Schreckenberg M (1992) A cellular automaton model for freeway traffc. J Phys 2:2221–2229

12. Neumann JV (1966) Theory of self-reproducing automata. University of Illinois Press, Champaign

13. Papadias D, Zhang J, Mamoulis N, Tao Y (2003) Query processing in spatial network databases. In: Proceedings of the 29th international conference on very large data bases (VLDB 2003), Berlin, pp 790–801

14. Shababi C, Kolahdouzan MR, Sharifzadeh M (2003) A road network embedding technique for K-nearest neighbor search in moving objects databases. GeoInformatica 7(3):255–273

15. Sistla P, Wolfson O, Chamberlain S, Dao S (1997) Modeling and querying moving objects. In: Proceedings of the 13th international conference on data engineering (ICDE 1997), Birmingham, pp 422–432

16. Speicys L, Jensen CS, Kligys A (2003) Computational data modeling for network-constrained moving objects. In: Proceedings of the 7th ACM international symposium on advances in geographic information systems (GIS 2003), New Orleans, pp 118–125

17. Su J, Xu H, Ibarra O (2001) Moving objects: logical relationships and queries. In: Proceedings of the 7th international symposium on spatial and temporal databases (SSTD 2001), Redondo Beach, pp 3–19

18. Ulam S (1972) Some ideas and prospects in biomathematics. Annu Rev Biophys Bioeng 1:272–292

19. Vazirgiannis M, Wolfson O (2001) A spatio-temporal model and language for moving objects on road networks. In: Proceedings of the 7th international symposium on spatial and temporal databases (SSTD 2001), Redondo Beach, pp 20–35

20. Wolfson O, Xu B, Chamberlain S, Jiang L (1998) Moving object databases: issues and solutions. In: Proceedings of the 10th international conference on scientific and statistical database management (SSDBM 1998), Capri, pp 111–122

Chapter 3
Moving Objects Tracking

Abstract The moving objects tracking system aims to monitor the locations of a set of objects which are traveling in a certain space, such as animals in fields and cars in road networks. It is a popular problem due to the importance in various application scenarios. In a typical moving objects application, large numbers of geographical positions of moving objects can be sampled by sensors or GPS following certain strategies, e.g., periodically, then sent from moving clients to the server and stored in a database. Therefore, continuously maintaining the current locations of moving objects in databases by proper tracking strategy becomes very important. The key problem is to reduce the location updates required to guarantee the error bound between an object's actual location and its current location in the tracking system, to provide precise results for locations query. In this chapter, we will introduce some typical researches on moving object tracking. Then, we introduce three representative network-constrained location update strategies (Net-LUM, ANLUM, and EuNetMOD), which can achieve better performances in terms of communication costs and location tracking accuracy.

Keywords Moving object tracking • Location updating • Traffic-adaptive location update • Traffic road network • Moving object databases

3.1 Introduction

Continuously maintaining current locations of moving objects in a database has become a fundamental issue nowadays because of its various applications, e.g., traffic management, logistics, taxi control, etc. As a result, moving object tracking has attracted great attention in recent years. In many LBS service systems, we hope to provide precise location of moving objects at any query time, but only some of the location updates of moving object can be preserved in central servers due to the limit of wireless bandwidth and the I/O of servers. The key problem is to find a proper balance between update cost and query precision.

X. Meng et al., *Moving Objects Management: Models, Techniques and Applications*, DOI 10.1007/978-3-642-38276-5_3,
© Tsinghua University Press, Beijing and Springer-Verlag Berlin Heidelberg 2014

The number of updates from moving objects to the server database depends on both the update frequency and the number of tracked objects. To reduce the location updates, most existing studies propose to lower the update frequency by a prediction method [3, 24, 25]. They usually use the model-based prediction which represents objects locations by some mathematical formulas based on their recent movements. The objects do not update their locations to server unless their actual positions deviate from the predicted positions by a given threshold. This provides a general principle for the location update policies in the moving objects database. In this chapter, we will introduce some typical location update strategies following this principle and three improved ones that have better performance in terms of tracking precision and update cost.

3.2 Representative Location Update Policies

So far, the research on tracking of moving objects has mainly focused on location update policies. Existing methods can be classified into four categories: threshold-based location updating, motion vector-based location updating, group-based location updating, and network-constrained location updating.

3.2.1 Threshold-Based Location Updating

In most real-world applications, the mostly used location tracking mechanisms include Fixed-Time Location Update Mechanisms (FTLU) and Fixed-Distance Location Update Mechanisms (FDLU) because of their simplicity. In FTLU and FDLU, moving objects update their position to central server at regular time or distance intervals. To improve efficiency and accuracy of tracking, Wolfson et al. [24] proposed the dead-reckoning update policies to reduce the update cost. The dead-reckoning update policies have three parts according to the threshold, namely, Speed Dead Reckoning (SDR) having a fixed threshold for all location updates, Adaptive Dead Reckoning (ADR) having different thresholds for different location updates, and Disconnection Detection Reckoning (DDR) with continuously decreasing threshold since the last location update. In [10], Gowrisankar and Nittel proposed a dead-reckoning policy that uses angular and linear deviations. Both of the two works assume that the destination and motion plan of the moving objects is known a priori. That means, the route traveled by moving objects is fixed and known. Lam et al. further proposed two location update mechanisms considering the effect of the continuous query results on the threshold [16]. The idea is that moving objects covered by the answers of queries have a lower threshold, leading to a higher location accuracy. Zhou et al. [26] also consider the precision of query results as a result of a negotiated threshold by the Aqua location updating scheme that they proposed.

3.2.2 Motion Vector-Based Location Updating

Motion vector-based location updating (MVLU) assumes the movements of an object can be represented with a *motion vector* $o(t)$, which can return the location of the object at any given future time stamp t. The motion vectors can be divided into two types: *linear* models [20, 25] and *nonlinear* models [1, 22]. Given location $o(t_0)$ of an object o' and its velocity v_0 at time t_0, the linear models estimate the object's position at time t by the formula $o(t) = o(t_0) + v_0 \times (t - t_0)$. The nonlinear models capture the object's movements by more sophisticated mathematical formulas. For example, Aggarwal and Agrawal [1] uses a quadratic function $o(t) = o(t_0) + v_0 \times (t - t_0) + a_0 \times (t - t_0)^2 / 2$, where a_0 is the current acceleration profile of o. Paper [21] introduces the *recursive motion function* (RMF) method to support both types of motion functions for accuracy improvement. By performing location prediction at both front side (moving objects) and back side (server) with the same motion function, the time of location updates can be reduced. Clearly, this method relies on an assumption that historical motion pattern of an object will continue in the future, but it is often true only for a near future. In reality, a group of them may show common motion patterns for a period of time. To further reduce the location updates, this observation motivates the group-based tracking.

3.2.3 Group-Based Location Updating

With the emergence of short-range communication (e.g., mobile P2P) supporting effective grouping together of individual objects, group-based location update mechanism that attempts to reduce the update cost of large-scale moving objects has become increasingly popular. The idea of partitioning objects nearby into coarse-grained groups as the basic tracking units is widely accepted especially in the area of Personal Communication Networks (PCS) [12, 18]. The early concept of group-based location updating can be referred to [15, 23]. Objects in the same group are assumed to move identically as a particular object elected to represent others. For the elected object in each group, the group location can be predicted with the motion vector-based tracking techniques. Since the objects are moving at their own discretion, the grouping is dynamic and it is done by online group self-organizing with local view in [2, 15]. The principal concept of group is a set of objects moving closely and similarly. By applying moving object clustering on trajectory data, we can achieve even better dynamic grouping with consideration of historical patterns. That is, moving objects in the same group are not only *crowding closely enough* in spatial aspect but also *accompanying long enough* with each other in temporal aspect. These works are *moving cluster* [13], *flock* [11], *convoy* [14], etc.

3.2.4 Network-Constrained Location Updating

More recently, increasing research interests are focused on network-constrained moving objects tracking [4, 9, 17, 19], in which the network distance is used instead of Euclidean distance. Since moving objects are constrained in the networks, it is possible to use the spatial and topology features of road network to facilitate moving object tracking. In [4], Civilis and Jensen et al. proposed a road-network-based location tracking mechanism for moving objects based on the speed patterns and acceleration profiles of road segments, so as to reduce the time of location updates. Such mechanism could be quite effective, but its performance heavily depends on GPS logs and accelerating profiles, which limits its usability in some real-world applications. In [6], Ding and Güting have proposed an MODTN model and provided some rough location update principles for network-constrained moving objects. And later, it is also noted that they have achieved a series of novel location update mechanisms based on the proposed data model, such as the essentially segment-based tracking method (Net-LUM) [9], the MVLU/FDLU-adaptive tracking method (ANLUM) [8], and the "mobile-map-free" tracking method (EuNetMOD) [5].

As most of the LBS services target to the scenarios where objects move on a predefined network like road network, we will particularly introduce three representative moving object tracking strategies (Net-LUM, ANLUM, and EuNetMOD) in the following sections of this chapter.

3.3 Network-Constrained Moving Objects Modeling and Tracking

Location update strategy is one of the most important factors that affect the performance of moving objects databases. In this section, a new location update mechanism, Location Update Mechanism for Network-Constrained Moving Objects (Net-LUM), is proposed. Through active motion vector-based network matching and special treatment with junctions, Net-LUM can achieve better performances in terms of communication costs and location tracking accuracy.

3.3.1 Data Model for Network-Constrained Moving Objects

In this section, we present the data model for network-constrained moving objects. The model is an improvement to the MODTN model proposed in [6]. The main improvements are as follows: (1) The geometry of a junction is expressed by a point plus a radius so that it is considered as an area instead of as a point; (2) a "graph point" value can be expressed either by a (*rid, pos*) pair or by a junction

ID, to accommodate the situation when the moving object is inside a junction area; and (3) the "motion vector" and the "moving graph point" definitions are extended accordingly to meet the situation when the moving object is inside a junction. Let $junct(jid)$ and $route(rid)$ be functions which return the junction and the route corresponding to the specified identifiers, respectively.

Definition 3.1 (Graph). A transportation graph (or graph) G is defined as a pair:

$$G = (Routes, Juncts)$$

where $Routes$ is a set of routes and $juncts$ is a set of junctions.

Definition 3.2 (Route). A route of graph G, denote by r, is defined as follows:

$$r = (rid, geo, len, fd)$$

where rid is the identifier of r, geo is a polygon line (or polyline) which describes the geometry of r (the beginning point and the end point of geo are called "0-end" and "1-end," respectively), len is the length of r, and $fd \in \{+, -, \pm\}$ is the traffic flow directions allowed in r.

Definition 3.3 (Junction). A junction of graph G, denoted by j, is defined as follows:

$$j = (jid, loc, ((rid_i, pos_i))_{i=1}^n, \gamma, m)$$

where jid is the identifier of j, loc is the location of j which can be presented as a point value in the $X \times Y$ plane, $((rid_i, pos_i))_{i=1}^n$ describes the routes connected by j, γ is the radius of the junction area, and m is the connectivity matrix [6] of j.

The radius γ can describe the size of the junction. That is to say, the junction is viewed as a junction area instead of a point in this model.

Definition 3.4 (Graph Point). A graph point is a point residing in the graph. The position of a graph point gp can have two possibilities: it is either in a junction (called "in junction") or on a route (called "in route"). For every route, we suppose that its total length is 1, so that any location inside the route can be presented by a real number $p \in [0, 1]$. We also define two Boolean functions $IsinJunct(gp)$ and $IsinRoute(gp)$ to check whether gp is in junction or in route.

The dynamic position of a network-constrained moving object is modeled as a moving graph point, which is a function from time to graph point. Discretely, a moving graph point is expressed as a sequence of motion vectors, and each motion vector describes the movement of the moving object at a certain time instant.

Definition 3.5 (Motion Vector). A motion vector, mv, is a snapshot of the moving object's movement and is generated by location updates. mv is defined as follows:

$$mv = (t, gp, \vec{v})$$

where t is a time instant, $gp \in GP$ is a graph point describing the location of the moving object at time t, and $\vec{v}$ is the speed measure of the moving object at time t. $\vec{v}$ is a real number value representing the current speed (its absolute value) and direction (its sign) if gp is on the route and set as $\perp$ ($\perp$ means "undefined") if gp is in a junction.

Definition 3.6 (Moving Graph Point). A moving graph point mgp is defined as

$$dmgp = (mv_i)_{i=1}^{n}$$

where $mv_i = (t_i, gp_i, \vec{v}_i)(1 \leq i \leq n)$ is the ith motion vector of the moving graph point, and for $\forall i \in \{1, \ldots, n-1\}$, we have: $t_i < t_{i+1}$.

For a running moving object mo, its last motion vector, $mv_n = (t_n, gp_n, \vec{v}_n)$, contains key information for prediction and location updates, and we call it "active motion vector." Through the active motion vector (i.e., the latest one), we can derive the estimated location of the moving object.

3.3.2 Location Update Strategies for Network-Constrained Moving Objects

In this section, we propose Location Update Mechanism for Network-Constrained Moving Objects (Net-LUM). The basic idea of Net-LUM is the "Inertia Principle," which assumes that the moving object will continue to move along the current route at roughly steady speed for some more time. Whenever this assumption becomes invalid, the moving object will launch a location update so that the up-to-date information of the moving object can be reported to the database server.

Moving objects continuously compare their current moving parameters (e.g., route identifier, location, speed, and direction) with the active motion vector it has submitted at last location update. The key problem of tracking is to find proper criteria for decision making on when a new location update should be made with both high precision and low cost guarantee.

In Net-LUM, we define three kinds of location updates, IDTLU, DTTLU, and STTLU. Each of them corresponds to an application domain, and they work together to fulfill effective location tracking of moving objects. Generally, IDTLU and DTTLU are basic strategies, while STTLU is optimal and only needed for uncertainty management [7].

Definition 3.7 (ID-Triggered Location Update (IDTLU)). For a running moving object, whenever it transfers from one route to another, a location update will be triggered to reflect the change of route identifiers. We call this kind of location updates ID-Triggered Location Updates (IDTLU).

Definition 3.8 (Distance-Threshold-Triggered Location Update (DTTLU)). When the moving object mo is running along a certain route, it repeatedly compares its actual position (denoted as gp_{gps}) with the computed position derived from the active motion vector (denoted as gp_{cmp}). If one of the following two conditions is met, (1) the distance between gp_{gps} and gp_{cmp} exceeds a threshold ξ and (2) gp_{cmp} is in junction while gp_{gps} is in the active route, a new location update will be triggered to report the actual location of the moving object. This kind of location updates is called Distance-Threshold-Triggered Location Updates (DTTLU).

Let us consider how to derive the estimated location gp_{cmp} from the active motion vector $mv_n = (t_n, gp_n, \overrightarrow{v}_n)$. If gp_n is in route (suppose $gp_n = (rid_n, pos_n)$), then the estimated position at the current time t_{now} is $gp_{cmp} = (rid_n, pos_{cmp})$, where pos_{cmp} can be computed with the following formula:

$$pos_{now} = pos_n + \frac{vm_n \times (t_{now} - t_n)}{route(rid_n).length}$$

Otherwise, if gp_n is in junction (suppose $gp_n = jid_n$), the estimated position at time t_{now} is still in the junction, and therefore we have $gp_{cmp} = jid_n$. A special case of position estimation that should be considered is that when the moving object is near the end of a route and the actual speed is lower than $|\overrightarrow{v}_n|$ (i.e., the absolute value of $\overrightarrow{v}_n$). In such case, the estimated position can exceed the scope of $[0, 1]$ and we can interpret the extra value as the distance covered by the moving object in other routes after it finishes the current route, so that the location update policy does not need to be changed.

Definition 3.9 (Speed-Threshold-Triggered Location Update (STTLU)). Suppose that the active motion vector of the moving object mo is $mv_n = (t_n, gp_n, \overrightarrow{v}_n)$. If $\overrightarrow{v}_n$ is defined, then mo will repeatedly compare its actual speed measure $\overrightarrow{v}_{gps}$ with $\overrightarrow{v}_n$ during its move. Location update is triggered if one of the following two conditions is met: (1) The difference between $\overrightarrow{v}_{gps}$ and $\overrightarrow{v}_n$ exceeds a threshold ψ; and (2) $\overrightarrow{v}_{gps}$ and $\overrightarrow{v}_n$ are in different directions. We call this update mechanism Speed-Threshold-Triggered Location Updates (STTLU).

STTLU ensures that the speed of the moving object is between $(|\overrightarrow{v}_i| - \psi)$ and $(|\overrightarrow{v}_i| + \psi)$ for any two consecutive location updates with motion vector $mv_i = (t_i, gp_i, \overrightarrow{v}_i)$ and $mv_{i+i} = (t_{i+i}, gp_{i+i}, \overrightarrow{v}_{i+i})$ if gp_i is in route.

Based on the above definitions, we can have the following two important inferences (proofs omitted).

Theorem 3.1. *Location update can happen at most one time inside a junction area.*

Theorem 3.2. *If moving object mo triggers a location update inside a junction, then when it drives out of the junction, it will update another location immediately.*

From the above analysis, we can see that Net-LUM can dramatically reduce location update costs around junctions. In real-world traffic systems, moving objects often run most irregularly around junctions. If junctions are not treated separately, there would be a lot of location updates, as illustrated in Fig. 3.1a. In comparison based on Net-LUM-based tracking as Fig. 3.1b, much less updates around junction

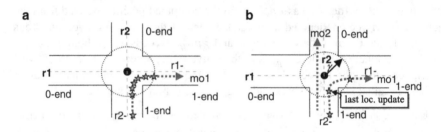

Fig. 3.1 Location updates around a junction. (**a**) Location updates without treating junctions separately. (**b**) Location updates around a junction in Net-LUM

($mo1$) is required or even no update is needed if the moving object passes junction ($mo2$) in a roughly steady speed. As an ideal situation, moving object may drive through the whole route without any location update, even though it passes through multiple junctions.

In Net-LUM, IDTLU, DTTLU, and STTLU work together to provide a complete location tracking mechanism for moving objects. The overall location update algorithm is shown in Algorithm 1 (we suppose that the algorithm is called frequently enough so that key location update chances will not be missed). In Algorithm 1, the function *TransformtoRoute*(gp, $gp*$, rid) transforms a graph point value from the *jid* form to the (rid, pos) form inside *route*(rid); *CreateLUM*() and *SendtoSVR*() create and send a location update message, respectively.

3.4   A Traffic-Adaptive Location Update Mechanism

Traffic in big city like New York is usually heavy especially in rush hours. In traffic jams, moving objects have to frequently speed up and gear down for a long time, which means that intensive location updates should be triggered in a short distance. In this section, we propose Adaptive Network-constrained moving object Location Update Mechanism (ANLUM) to solve this problem. In ANLUM, the moving object can switch between different location tracking policies (e.g., MVLU and FDLU) based on the traffic condition. When the moving object is not inside traffic jams, the MVLU method is used. If the moving object encounters a traffic jam, we switch to the distance based on FDLU tracking. To well balance different policies, the key problem is to detect traffic jam accurately and effectively.

Automatic traffic jam detection in real time is a necessary but challenging problem. Under the general ANLUM principle, we have two alternatives: Autonomic ANLUM (ANLUM-A), which detects traffic jams on the moving object itself, and Centralized ANLUM (ANLUM-C), which detects traffic jams on the server side.

Algorithm 1: $Net\_LUM_{mo}(G; \xi; \psi;)$ //Algorithm running at the moving object end

---

Input: the graph of the traffic network G, the distance threshold ξ, and the speed threshold ψ
begin

> **while** *mo is active* **do**
>
>> Read GPS signal, and get Euclidean position $p = (x, y)$, speed v, and direction d;;
>> Transform (p, d, gp, fd); //network-matching; $\overrightarrow{v} \leftarrow fd * v$;
>> $mv_n \leftarrow (t_n, gp_n, \overrightarrow{v}_n)$ the active motion vector sent at the last location update;;
>> $actv\_rid \leftarrow$ be the ID of the active route; **if** $IsinJunct(gp_n)$ **then**
>>
>>> **if NOT** $(IsinJunct(gp)$ and $(gp = gp_n))$ **then**
>>> | SendtoSVR(CreateLUM($mid, t_{now}, gp, \perp$)); //DTTLU or IDTLU;
>>> **end**
>>
>> **else**
>>
>>> **if** $IsinJunct(gp)$ **then**
>>> | TransformtoRoute($gp, gp\_inactvroute, actv\_rid$);;
>>> **else**
>>> | $gp_inactvroute \leftarrow gp$;
>>> **end**
>>> // $gp\_inactvroute$ is in route. suppose $gp\_inactvroute = (rid, pos)$; **if** $rid \neq actv\_rid$ **then**
>>> | SendtoSVR(CreateLUM($mid, t_{now}, gp, \overrightarrow{v}$)); //IDTLU
>>> **else**
>>> | $gp_{cmp} \leftarrow$ the computed position derived from mv_n; **if** *one of the DTTLU conditions is met* **then**
>>> | | SendtoSVR(CreateLUM($mid, t_{now}, gp, \overrightarrow{v}$)); //DTTLU
>>> | **else if** *one of the STTLU conditions is met* **then**
>>> | | SendtoSVR(CreateLUM($mid, t_{now}, gp, \overrightarrow{v}$)); //STTLU
>>> | **end**
>>> **end**
>>
>> **end**
>
> **end**

end

Fig. 3.2 Route-based transportation network model

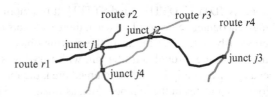

In ANLUM, we use the route-based model to represent the transportation network. A network N is modeled as the form Definition 3.1. Let *junct(jid)* and *route(rid)* be functions which return the junction and the route corresponding to the specified identifiers, respectively. From Fig. 3.2 we can see that two or more routes can intersect each other either at their end points or at the middle points, so that a junction can appear at any position of a route.

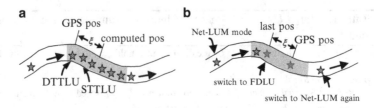

Fig. 3.3 Location updates in traffic jam with Net-LUM and ANLUM-A. (**a**) Net-LUM.
(**b**) Reduced location updates with ANLUM-A

A network position $npos \in D(N)$ is a point residing in N, whose position have
two possibilities: either in route or in junction. In the former case, a real number
$p \in [0, 1]$ has to be indicated to further describe its position inside the route.

For a network-constrained moving object mo, its trajectory $mo.traj$ is a function
from time to network position, that is, $f : T \rightarrow D(N)$, where T is the domain
of time instants. In implementation, the function f has to be translated to a
discrete form represented as a sequence of motion vectors. A motion vector mv
is defined as $mv = (t, npos, v, d)$, where t is a time instance, $npos$ is a network
position, v is the speed, and d is moving direction of the object. The trajectory
of a network-constrained moving object mo is defined as $traj = (mv_i)_{i=1}^{n}$ where
$mv_i = (t_i, npos_i, v_i, d_i)(1 \leq i \leq n)$ is the ith motion vector of mo.

In the following, we introduce ANLUM-A and ANLUM-C methods in detail and
discuss how they collaborate in the ANLUM mechanism.

3.4.1 The Autonomic ANLUM (ANLUM-A) Method

In [6, 9], we have defined an MVLU-based mechanism, Net-LUM, for network-
constrained moving objects. Net-LUM basically consists of three kinds of location
updates, IDTLU, DTTLU, and STTLU. It is a motion vector-based approach, and it
may encounter some problem when there exist lots of traffic jams in the network, as
previously stated. For instance, the moving object can spend a long period of time
before driving through a traffic jam with very low speed. Since the moving pattern
changes frequently inside the jammed area, the moving object can trigger multiple
location updates in the jammed area, as illustrated in Fig. 3.3a.

To reduce location updates in traffic jams, we propose the Autonomic ANLUM
(ANLUM-A) method as follows. Given a new location update, it continuously
computes the average speed between the last location update time and the current
time, denoted as $\overrightarrow{v}$. The update will work in the distance-based FDLU mode if $\overrightarrow{v}$ is
slower than a predefined slow speed threshold v_{slow} and work in the Net-LUM mode
otherwise. In this way, the overall location update cost can be reduced, as illustrated
in Fig. 3.3b. The detailed ANLUM-A algorithm, running on the moving object side,
is shown in Algorithm 2.

Algorithm 2: ANLUM-A$(N; v_{slow})$ //The ANLUM-A algorithm

Input: the transportation network N, the slow speed threshold v_{slow}

begin

     mode $\leftarrow$ Net-LUM;;

     while mo *is active* **do**

         Read GPS signal, and get Euclidean position $p = (x, y)$, speed v, and direction d;

         Transform $p \rightarrow npos$;

         $mv_{now} \leftarrow (t_{now}, npos, v, d)$;

         $mv_n \leftarrow (t_n, npos_n, v_n, d_n)$; //the active motion vector sent at the last location update;

         if *(mode = Net-LUM)* **AND** *(Net-LUM(mv_n, mv_{now})=True)* **then**

             if *AverageSpeed(mv_n, mv_{now})* $< v_{slow}$ **then**

                 mode $\leftarrow$ FDLU;

                 SendtoSVR(CreateLUM($mid, t_{now}, npos, 0, \perp$));

             else

                 mode $\leftarrow$ Net-LUM;

                 SendtoSVR(CreateLUM($mid, t_{now}, npos, v, d$));

             end

         end

         if *(mode = FDLU)* **AND** *(Distance($npos_n, npos$) $> \xi$)* **then**

             if *AverageSpeed(mv_n, mv_{now})* $< v_{slow}$ **then**

                 mode $\leftarrow$ FDLU;

                 SendtoSVR(CreateLUM($mid, t_{now}, npos, 0, \perp$));

             else

                 mode $\leftarrow$ Net-LUM;

                 SendtoSVR(CreateLUM($mid, t_{now}, npos, v, d$));

             end

         end

     end

end

In Algorithm 2, the function Net-LUM(mv_n, mv_{now}) selects the best policy from (IDTLU, DTTLU, and STTLU) according to the current condition based on the rules stated in previous section. Function *Distance()* returns the distance between two network positions, functions *CreateLUM()* and *SendtoSVR()* generate the location update message and send it to the server respectively, and function *AverageSpeed()* derives the average speed $\vec{v}$ from mv_n, mv_{now} as follows:

$$\vec{v} = \frac{distance(npos, npos_n)}{t_{now} - t_n}$$

This algorithm works on the moving object side, and it notifies the server about its mode through the parameters v and d. If v and d are set to 0 and $\perp$ (undefined), respectively, then the moving object is in the FDLU mode. Otherwise, if v and d assume real values sampled from GPS, the moving object is in the Net-LUM mode. In this way, the server and the moving object are synchronized so that the computed position can be derived correctly at the server.

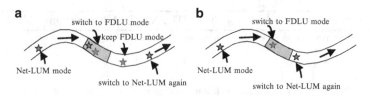

Fig. 3.4 Location updates in traffic jam with ANLUM-A and ANLUM-C. (**a**) ANLUM-A. (**b**) ANLUM-C

3.4.2 The Centralized ANLUM (ANLUM-C) Method

The above ANLUM-A method can reduce location updates inside traffic jams to some extent. However, there are still some drawbacks with this method. For instance, moving objects may mistakenly detect that it is inside a traffic jam when it only slows down temporarily. Besides, it needs at least two location updates to switch from MVLU to FDLU or vice versa so that the location update cost could still be unnecessarily high in some cases, especially when it drives through small-sized traffic jams, as shown in Fig. 3.4a.

To solve this problem, we further propose an improved ANLUM method, the Centralized ANLUM method (ANLUM-C), which detects traffic jams on the server side by analyzing the recent trajectory data of moving objects. Detected traffic jam status is broadcasted to moving objects in real time, so that moving objects can know whether they are inside traffic jams and use proper tracking mode accordingly. With such global jam information available, the update cost around jams can be further reduced, as shown in Fig. 3.4b.

The problem is how to compute the jammed areas of a certain route r from the trajectory data. A jammed area is defined as a continuous part of the route where all moving objects pass through with speed slower than the predefined slow speed threshold v_{slow} in the last Δt time (Δt is called "statistic window" whose duration could be 5–10 min). Multiple jammed areas could coexist in r.

The basic idea of the jam detection in ANLUM-C method is to compute the slow segment of each moving object from the corresponding trajectory first, then to compute the intersection of all moving objects' slow segments, so that the jammed areas of r can be found. In this procedure we should consider the "driving range" of each moving object, which is the route segment where the moving object has driven through. Since the moving object can only have a voice for its driving range, we modify the intersection operator with driving range considered.

Suppose that $traj = (t_i, npos_i, v_i, d_i)_{i=1}^n$ is an arbitrate trajectory. Function $SlowSegRange(traj, r, fd, \Delta t, v_{slow})$ returns a pair (δ, φ), where $\delta \subseteq [0, 1]$ is a segment of r through which the corresponding moving object drives along the direction fd with a speed slower than a given threshold v_{slow} and $\varphi \subseteq [0, 1]$ is a segment of r that the moving object has driven through along the direction fd. We call the pair (δ, φ) "Slow Segment with Range (SSR)." SSR can be computed easily

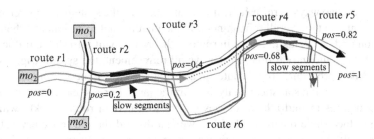

Fig. 3.5 Trajectories and the corresponding slow speed segments and driving ranges

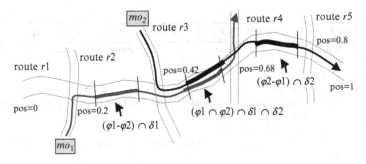

Fig. 3.6 Semantics of the $\overline{\cap}$ operator

from *traj* since the average speed between any two adjacent updates can be derived from the corresponding motion vectors. Figure 3.5 shows some example trajectories and the derived SSR of three moving objects.

Based on SSR, we define an "intersection with range" operator $\overline{\cap}$, which works very similar to the intersection operator $\cap$ except that the driving range is considered. Suppose that $A = (\delta 1, \varphi 1)$ and $B = (\delta 2, \varphi 2)$ are two arbitrary SSRs. The semantics of the $\overline{\cap}$, operator is defined as follows:

$$A \overline{\cap} B = (\delta 1 \cap \delta 2) \cup ((\varphi 1 - \varphi 2) \cap \delta 1) \cup ((\varphi 2 - \varphi 1) \cap \delta 2)$$

The $\overline{\cap}$ operator is designed for detecting the jammed area from trajectories. Suppose we have only two moving objects mo_1 and mo_2. The jammed area can have three possibilities: (1) the part of the route where both mo_1 and mo_2 have driven through with slow speed, that is, $\delta 1 \cap \delta 2$; (2) the part of the route where only mo_1 has driven through with slow speed while mo_2 does not drive through, that is, $(\varphi 1 - \varphi 2) \cap \delta 1$; and (3) the part of the route where only mo_2 has driven through but mo_1 does not touch, that is, $(\varphi 2 - \varphi 1) \cap \delta 2$, as shown in Fig. 3.6.

Suppose that $\Psi = traj_1, traj_2, \ldots, traj_n$ is the set of trajectories in the MOD system. Then for route r along the fd direction, its traffic jams can be computed as follows:

$$\alpha = \overline{\cap}_{i=1}^{n}(SlowSegRange(traj_i, fd, r, \Delta t, v_{slow}))$$

There are two ways to refresh the traffic jam status of the system: the incremental refreshment and the overall refreshment. In the incremental refreshment, whenever a location update occurs in a certain route r, the system will refresh the traffic jam status of the route in real time. In the overall refreshment, the system triggers the jam detection procedure for all routes in a periodical way. Some techniques can be used to speed up the statistical analysis of traffic jams. For instance, we can utilize index structures to find all trajectories related to a certain route quickly without traversing the whole database. Besides, we can discard the old trajectory data or keep them elsewhere so that the statistics can be conducted only on recent trajectory information which has much smaller size.

Algorithm 3: ANLUM-C(N; ξ; Jam)   //The ANLUM-C algorithm

Input: the traffic network N, the distance threshold ξ, the traffic jams of the system Jam
begin

    mode $\leftarrow$ Net-LUM;

    while *mo is active* **do**

        Read GPS signal, and get Euclidean position $p = (x, y)$, speed v, and direction d;

        Transform $p \rightarrow npos$;

        $mv_{now} \leftarrow (t_{now}, npos, v, d)$;

        $mv_n \leftarrow (t_n, npos_n, v_n, d_n)$; //the active motion vector sent at the last location update;

        if *(mode = Net-LUM) AND (Net-LUM(mv_n, mv_{now})=True)* **then**

            if *inside($npos$, Jam)* **then**

                mode $\leftarrow$ FDLU;

                SendtoSVR(CreateLUM(mid, t_{now}, $npos$, 0, $\perp$));

            else

                mode $\leftarrow$ Net-LUM;

                SendtoSVR(CreateLUM(mid, t_{now}, $npos$, v, d));

            end

        end

        if *(mode = FDLU) AND (Distance($npos_n$, $npos$) > ξ)* **then**

            if *inside($npos$, Jam)* **then**

                mode $\leftarrow$ FDLU;

                SendtoSVR(CreateLUM(mid, t_{now}, $npos$, 0, $\perp$));

            else

                mode $\leftarrow$ Net-LUM;

                SendtoSVR(CreateLUM(mid, t_{now}, $npos$, v, d));

            end

        end

    end

end

The result of the traffic jam detection is broadcasted to the moving objects so that any moving object knows its nearby traffic condition. Based on the traffic jam information, the moving object can decide which location update mode to be utilized. The detailed ANLUM-C algorithm running at the moving object side is given in Algorithm 3. In Algorithm 3, the moving objects switch between Net-LUM and FDLU according to whether it is located in a jammed area. The FDLU mode will be used if it is true, and otherwise it will switch to the Net-LUM mode.

3.5   A Hybrid Network-Constrained Location Update Mechanism

To track network-matched trajectories of moving objects is important in a lot of applications such as trajectory-based traffic-flow analysis and trajectory data mining. However, current network-based location tracking methods for moving objects need digital maps installed at the moving object side, which is not realistic for many scenarios. In this section, we briefly introduce a new framework, Euclidean batch sampling and Network-matched trajectory-based Moving Objects Database (EuNetMOD) model, to support network-matched trajectory tracking without digital maps installed at the moving object side.

In EuNetMOD, moving objects can read motion vectors at any time and send them to the server repeatedly. We call the operations of "reading motion vectors" and "sending them to server" as "sampling" and "location update," respectively.

For a certain moving object mo, its locations can be tracked by using a "Fixed-Time Sampling plus Fixed-Time Location Update (FTS + FTLU)" method. That is, in every τ_s time (say in every 15 s), mo samples a Euclidean-based motion vector of the form $(t, (x, y), v, d)$, where t is a time stamp and (x, y), v, d are the location, the speed, and the direction of mo at time t, respectively. Besides, every time when mo's direction change or speed change exceeds certain predefined thresholds, mo will trigger an extra sampling. The sampled motion vectors are temporarily kept in the local storage of the moving object, and in every τ_u time (say in every 3 min), mo sends the sampled motion vectors to the database server in batch.

Except the abovementioned "FTS + FTLU" location tracking strategy, we can also adopt other location update policies such as "FDS + FTLU" (Fixed-Distance Sampling plus Fixed-Time Location Update) or "FDS + FDLU" (Fixed Distance Sampling plus Fixed-Distance Location Update). The general rule is that mo should sample its Euclidean-based motion vectors relatively densely and sends the sampled data in batch to the server in relatively sparse time intervals.

When the database server receives a location update message which contains multiple Euclidean-based motion vectors, it will match the Euclidean-based motion vectors to the network so that we can get network-matched motion vectors of the form $(t, (x, y), v, d, npos)$, where t, (x, y), v, and d are from the original Euclidean-based motion vector and $npos$ is the corresponding network position. After that, it will find a network path between any two neighboring network-matched motion vectors so that the network path for the whole trajectory becomes available. Then it will discard unimportant motion vectors which are implicitly inferable from its predecessor and successor. The returned trajectory is "network matched" since the path information is explicitly expressed in the trajectory, as

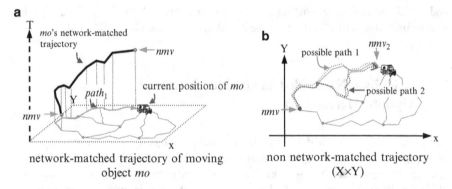

Fig. 3.7 Network-matched vs. non-network-matched trajectories

shown in Fig. 3.7a. Figure 3.7b shows a non-network-matched trajectory which can have ambiguity in deciding the path between two neighboring sampling points.

As stated above, since the path information (expressed as an edge sequence) is contained in the trajectory, unimportant motion vectors can be discarded if it is inferable from its neighboring motion vectors. In extreme cases, a network-matched trajectory can have only two network-matched motion vectors nmv_1, nmv_2 and the path between them (if the moving object runs in roughly steady speed), as illustrated in Fig. 3.7. Therefore, the storage of the network-matched trajectory can be optimized while the precision can still be reserved.

Each network-matched trajectory can describe a continuous movement of the moving object. To describe the dynamic locations of a moving object over a long period of time (say 3 months), multiple trajectories are needed. For simplicity, we still call the multiple trajectories of the same moving object as "trajectory" in this paper. In real-world applications, there exist situations when moving objects run outside the traffic network occasionally. In this case, EuNetMOD will keep it in the trajectory as its original Euclidean form. Besides, EuNetMOD allows network-matching to have multiple candidate network points coexisting so that the model can deal with more flexible situations.

3.6   Summary

This chapter introduces a few location update techniques to track network-constrained moving objects. On one hand, the techniques lower the location update frequency and ensure the accurate locations in tracking by a new prediction method and also minimize the cost of communication between tracking clients and tracking server. On the other hand, hybrid and adaptive location update strategies are presented according to movement features of objects moving on a Euclidean space and a road network, which further reduces the total number of location updates.

References

1. Aggarwal C, Agrawal D (2003) On nearest neighbor indexing of nonlinear trajectories. In: PODS, San Diego, 2003, pp 252–259
2. Chen J, Meng X, Li B, Lai C (2006) Tracking network-constrained moving objects with group updates. In: Proceedings of the 7th international conference on web-age information management (WAIM 2006), Hong Kong, pp 158–169
3. Civilis A, Jensen CS, Nenortaite J, Pakalnis S (2004) Efficient tracking of moving objects with precision guarantees. In: Proceedings of the 1st annual international conference on mobile and ubiquitous systems, networking and services, Cambridge, pp 164–173
4. Civilis A, Jensen CS, Pakalnis S (2005) Techniques for efficient road-network-based tracking of moving objects. IEEE Trans Knowl Data Eng 17(5):698–712
5. Ding Z, Deng K (2011) Collecting and managing network-matched trajectories of moving objects in databases. In: Proceedings of the 22nd international conference on database and expert systems applications (DEXA 2011), Toulouse, pp 270–279
6. Ding Z, Güting RH (2004) Managing moving objects on dynamic transportation networks. In: Proceedings of the 16th international conference on scientific and statistical database management (SSDBM 2004), Santorini Island, pp 287–296
7. Ding Z, Güting RH (2004) Uncertainty management for network constrained moving objects. In: Proceedings of the 15th international conference on database and expert systems applications (DEXA 2004), Zaragoza, pp 411–421
8. Ding Z, Guo L, Meng X (2009) Adaptive location update mechanism for network-constrained moving objects in changeful traffic conditions. In: Proceedings of the 10th international conference on mobile data management: systems, services and middleware (MDM 2009), Taipei, pp 417–423
9. Ding Z, Zhou X (2008) Location update strategies for network-constrained moving objects. In: Proceedings of the 14th international conference on database systems for advanced applications (DASFAA 2008), New Delhi, pp 644–652
10. Gowrisankar H, Nittel S (2002) Reducing uncertainty in location prediction of moving objects in road networks. In: Proceedings of the 2nd international conference on geographic information science (GIS 2002), Boulder, pp 228–242
11. Gudmundsson J, Kreveld M (2006) Computing longest duration flocks in trajectory data. In: Proceedings of the 14th ACM international symposium on geographic information systems (GIS 2006), Arlington, pp 35–42
12. Huh Y, Kim C (2002) Group-based location management scheme in personal communication networks. In: Proceedings of the international conference on information networking (ICOIN 2002), Cheju Island, pp 81–90
13. Jensen CS, Lin D, Ooi BC (2007) Continuous clustering of moving objects. IEEE Trans Knowl Data Eng 19(9):1161–1174
14. Jeung H, Yiu ML, Zhou X, Jensen CS, Shen H (2008) Discovery of convoys in trajectory databases. In: Proceedings of the 34th international conference on very large data bases (VLDB 2008), Auckland, pp 1068–1080
15. Lam GHK, Leong HV, Chan SC (2004) GBL: group-based location updating in mobile environment. In: Proceedings of the 9th international conference on database systems for advanced applications (DASFAA 2004), Jeju Island, pp 762–774
16. Lam KY, Ulusoy O, Lee TSH, Chan E, Li G (2001) An efficient method for generating location updates for processing of location-dependent continuous queries. In: Proceedings of the 6th international conference on database systems for advanced applications (DASFAA 2001), Hong Kong, pp 218–225
17. Li X, Han H, Lee J, Gonzalez H (2007) Traffic density-based discovery of hot routes in road networks. In: Proceedings of the 10th international symposium on spatial and temporal databases (SSTD 2007), Boston, pp 441–459

18. Mao Z, Douligeris C (2006) Group registration with local anchor for location tracking in mobile networks. IEEE Trans Mob Comput 5(5):583–595
19. Pfoser D, Jensen CS (2003) Indexing of network constrained moving objects. In: Proceedings of the 11th ACM international symposium on geographic information systems (GIS 2003), New Orleans, pp 25–32
20. Šaltenis S, Jensen CS, Leutenegger ST, Lopez MA (2003) Indexing the positions of continuously moving objects. SIGMOD Rec 29(2):331–342
21. Tao Y, Faloutsos C, Papadias D, Liu B (2004) Prediction and indexing of moving objects with unknown motion patterns. In: SIGMOD 2004, Paris, pp 611–622
22. Trajcevski G, Wolfson O, Xu B, Nelson P (2002) Real-time traffic updates in moving objects databases. In: Proceedings of the 13th international conference on database and expert systems applications (DEXA 2002), Aix-en-Provence, pp 698–704
23. Wang KH, Li B (2002) Efficient and guaranteed service coverage in partitionable mobile ad-hoc networks. In: Proceedings of the 21st international conference on computer communications (INFOCOM 2002), New York, pp 1089–1098
24. Wolfson O, Sistla AP, Camberlain S, Yesha Y (1999) Updating and querying databases that track mobile units. Distrib Parallel Databases 7(3):257–387
25. Wolfson O, Yin H (2003) Accuracy and resource consumption in tracking and location prediction. In: Proceedings of the 7th international symposium on spatial and temporal databases (SSTD 2003), Santorini Island, pp 325–343
26. Zhou J, Leong HV, Lu Q, Lee KC (2005) Aqua: an adaptive query-aware location updating scheme for mobile objects. In: Proceedings of the 11th international conference on database systems for advanced applications (DASFAA 2005), Beijing, pp 612–624

Chapter 4
Moving Objects Indexing

Abstract For querying large amounts of moving objects, a key problem is to create efficient indexing structures that make it possible to effectively answer various types of queries. Traditional spatial indexing approaches cannot be used because the locations of moving objects are highly dynamic, which leads to frequent updates of index structures, which in turn will cause huge overheads. In this chapter, we first introduce a few of representative indexing methods including the R-Tree, TPR-tree, STR-Tree, TB-tree, and MON-tree. Then, we propose two new index methods for moving objects, one for indexing frequently updated trajectories in spatial networks and another for indexing the past, present, and anticipated future positions of moving objects.

Keywords Spatial index • Spatio-temporal index • Index update • Trajectory • Spatial network • Moving object databases

4.1 Introduction

The trajectory data to be managed by moving objects databases are essentially massive dynamic data due to the large number of moving objects for manage and their frequent location updates. As a result, processing of queries with complex spatio-temporal constraints on trajectory data is usually time consuming, none to speak when the concurrency of queries is considered. To ensure that desired result data in the huge data set can be efficiently located, it is thus important to create effective index structures for performance guarantee.

Effective indexing on massive trajectory data is a challenging issue because of the intrinsic spatio-temporal features and frequent location updates of moving objects. They cause the fact that traditional indexing approaches cannot be applied: First, indexing on moving object trajectory data requires a 3-dimensional index structure in the $X \times Y \times T$ space, but it cannot be derived by simply adding a dimension to traditional 2-dimensional spatial index because of the complex feature interaction;

X. Meng et al., *Moving Objects Management: Models, Techniques and Applications*, DOI 10.1007/978-3-642-38276-5_4,
© Tsinghua University Press, Beijing and Springer-Verlag Berlin Heidelberg 2014

second, given that the location of an object keeps changing along its movement, it is not realistic to update the index once the moving object location is changed. Otherwise, index update would be so frequent that the index itself could become a bottleneck of the system.

To handle the challenges mentioned above, lots of efforts have been made so far to develop effective moving object indexes. In general, existing index methods for moving objects database can be divided into two categories: (1) indexing the historical trajectories and (2) indexing their current positions for prediction of future trajectories.

In historical trajectory indexes, typical indexing approaches include [12, 14, 17], all based on 3-D variations of R-Tree and R*-tree with a goal to minimize the storage and query processing cost. However, they still have too expensive update cost, which can be greatly reduced if the structure of underlying network is available for use (road network constraint movement scenarios). In recent years, more efforts have been made for trajectory data indexing on moving objects in spatial networks, with methods like dimension transformation-based indexes [13] and two-layered FNR-Tree [8] been proposed. In addition, the MON-tree approach [2] improves the performance of the FNR-tree by representing each edge by multiple line segments (i.e., polyline) instead of just one line segment. Also, the Spatio-Temporal R-Tree(STR-tree) [14] and the Trajectory-Bundle tree(TB-tree) [14] further consider the trajectory property as well as allow for typical range queries.

In current position indexes of moving objects, some early studies [1, 11] employ dual transformation techniques that represent the predicted positions as points moving in a two-dimensional (2D) space. Recent works are more focused on practical implementation, including the R-Trees-based TPR-tree [16] and its variations [15] and the B^+-tree-based B^x-tree [10]. But the update performance of above index mechanisms is not satisfactory. As an improvement, the PMR Quad-tree-based index is proposed in [9], which adopts a trajectory segment shared structure while depicting an efficient update algorithm. A dynamic data structure, called adaptive unit, is introduced in [4], which groups neighboring objects with similar movement patterns and captures the movement bounds of the objects based on traffic behavior to reduce updates. Also, a spatial index for the road network is then built over the adaptive unit structures, which forms the ANR-tree [5].

However, the update performance of all above index strategies is not good enough for many practical applications yet. A network-constrained Moving Object Sketched-Trajectory R-Tree(MOSTR-Tree) is proposed in [6] by Ding to handle this problem. It is robust for frequent updates because a coarser granularity trajectory named sketched trajectory instead of original trajectory is used. Also, because of the lacking of index model that can support queries on historical, present, and near-future possible locations, Ding et al. further developed a Network-constrained moving objects Dynamic Trajectory R-Tree(NDTR-Tree) in [7] where a hybrid structure with two layers (each layer based on R-Tree) is used to achieve this.

In this chapter, several typical moving object indexes in Euclidean space and in spatial networks are introduced first, specifically including R-Tree, TPR-tree,

Spatio-Temporal R-Tree, Trajectory-Bundle tree, and MON-tree. Afterwards, two advanced moving objects indexing methods (MOSTR-Tree and NDTR-Tree) are presented and discussed in detail.

4.2 Representative Indexing Methods

This section presents several representative indexing methods including the R-Tree [9], the TPR-tree [16], the Spatio-Temporal R-Tree [14], the Trajectory-Bundle tree [14], and MON-tree [2].

4.2.1 The R-Tree

The R-Tree [9] is proposed by Antomn Guttman, which is a height-balanced indexing structure as an extension of the B-tree in multidimensional space.

Each node of the R-Tree contains a hyper-rectangle in d dimensions. The rectangles of leaf nodes contain spatial objects indexed, and the rectangles of internal nodes cover all rectangles in the lower nodes. The boundaries of the rectangles are made as tight as possible. These rectangles are called minimum bounding rectangles or MBRs. An entry in a leaf node is of the form: (MBR_o, p_o), where MBR_o is the MBR of the indexed spatial object and p_o is a pointer to the actual object tuple in the database. An entry in an internal node is of the form: (MBR_c, p_c), where MBR_c is the MBR covering all MBRs in its child node and p_c is the pointer to its child node c. The number of entries in each R-Tree node, except for the root node, is between two specified parameters m and M ($m \leq M/2$). The parameter M is termed the fanout of the R-Tree. Unlike B-tree, the MBRs of nodes at the same level in an R-Tree are allowed to overlap. Hence, searching an object may involve traversing several paths in the R-Tree. When a node becomes overfull, it undergoes a split. Efficient heuristics and pruning are used to reduce the expected number of paths visited by subsequent searches.

Figure 4.1 represents the R-Tree corresponding to the spatial distribution of objects (solid rectangles in this case) below it.

The R-Tree has the following features:

- The R-Tree is height balanced. The root node has at least two children nodes, and all leave nodes are in the same level of the tree.
- If M is the maximum number of entries in an R-Tree, then $m \leq M/2$, where m is the minimum number of entries.
- Height of the R-Tree is $|\log_m N| - 1$.
- Maximum number of nodes is $\lceil N/m \rceil + \lceil N/m^2 \rceil + \cdots + 1$.

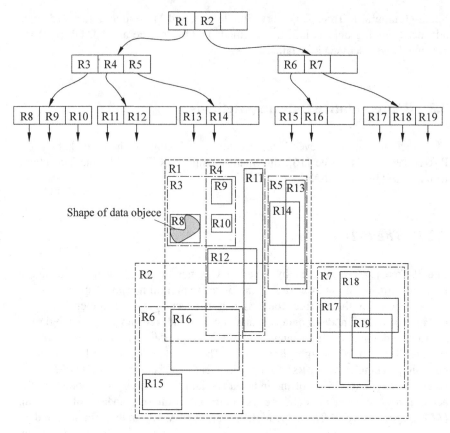

Fig. 4.1 Structure of the R-Tree

4.2.2 The TPR-Tree

The time-parameterized R-Tree(TPR-tree) [16] is proposed by Simonas Saltenis, which is a balanced, multi-way tree with the structure of the R-Tree. The TPR-Tree naturally extends the R*-tree [3] and efficiently indexes the current and anticipated future positions of moving point objects (or "moving points" for short).

Entries in leaf nodes are pairs of the position of a moving point and a pointer to the moving point, and entries in internal nodes are pairs of a pointer to a subtree and a rectangle that bounds the positions of all moving points or other bounding rectangles in that subtree.

In the TPR-Tree, a moving object o is represented by (1) an MBR o_R that denotes its extent at reference time 0 and (2) a velocity bounding rectangle(VBR) $o_V = \{o_{V1-}, o_{V1+}, o_{V2-}, o_{V2+}\}$, where $o_{Vi-}(o_{Vi+})$ describes the velocity of the lower (upper) boundary of o_R along the ith dimension ($1 \le i \le 2$).

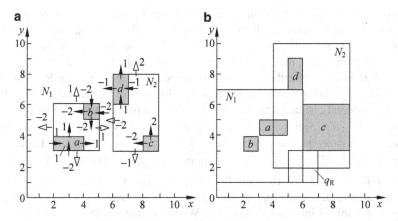

Fig. 4.2 Entry representations in a TPR-Tree. (**a**) MBRs and VBRs at time 0. (**b**) MBRs at time 1

Fig. 4.3 N_1 is tightened
during an insertion at time 1

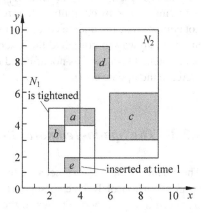

Figure 4.2a shows the MBRs and VBRs of four objects a, b, c, and d. The arrows
(numbers) denote the directions (values) of their velocities, where a negative value
implies that the velocity is toward the negative direction of an axis. The VBR of
a is $a_V = \{1, 1, 1, 1\}$ (the first two numbers are for the X dimension), while
those of b, c, and d are $b_V = \{-2, -2, -2, -2\}$, $c_V = \{-2, 0, 0, 2\}$, and $d_V =
\{-1, -1, 1, 1\}$, respectively. A non-leaf entry is also represented by an MBR and a
VBR. Specifically, the MBR (VBR) tightly bounds the MBRs (VBRs) of the entries
in its child node. In Fig. 4.2b, the objects are clustered into two leaf nodes N_1, N_2,
whose VBRs are $N_{1V} = \{-2, 1, -2, 1\}$ and $N_{2V} = \{-2, 0, -1, 2\}$ (their directions
are indicated using white arrows).

Figure 4.3 shows the MBRs at time stamp 1 (notice that each edge moves
according to its velocity). The MBR of a non-leaf entry always encloses those of
the objects in its subtree, but it is not necessarily tight. For example, N_1 (N_2) at
time stamp 1 is much larger than the tightest bounding rectangle for $a, b(c, d)$.
A predictive window query is answered in the same way as in the R*-tree, except
that it is compared with the (dynamically computed) MBRs at the query time.

For example, the query q_R at time stamp 1 in the figure visits both N_1 and N_2 (although it does not intersect them at time 0). The TPR-Tree is optimized for time stamp queries in interval $[T_C, T_C + H]$, where T_C is the current update time and H is a tree parameter called the horizon (i.e., how far the tree should "see" in the future). The update algorithms are exactly the same as those for the R*-tree and are obtained by simply replacing the four penalty metrics of the previous section with their integral counterparts. Specifically, the area (or perimeter) of an entry N equals $\int_{T_C}^{T_C+H} A(N,t)\mathrm{d}t$ (or $\int_{T_C}^{T_C+H} P(N,t)\mathrm{d}t$), where $A(N,t)$ (or $P(N,t)$) returns the area (perimeter) of N at time t. Similarly, the overlap (or the centroid distance) between two MBRs N_1 and N_2 is computed as $\int_{T_C}^{T_C+H} OVR(N_1, N_2, t)\mathrm{d}t$ (or $\int_{T_C}^{T_C+H} CDist(N_1, N_2, t)\mathrm{d}t$), where $OVR(N_1, N_2, t)$ (or $CDist(N_1, N_2, t)$) returns the overlapping area (centroid distance) between N_1 and N_2 at time t. These integrals are solved into closed formulas. When an object is inserted or removed, the TPR-Tree tightens the MBR of its parent node. Figure 4.3 shows the MBRs after inserting a new object e (into N_1) at time 1. N_1 is adjusted to the tightest MBR bounding a, b, e, by computing their respective extents at time 1. Note that this does not compromise the update cost because N_1 must be loaded (written back) from (to) the disk anyway to complete the insertion. On the other hand, the MBR of N_2 is not tightened because it is not affected by the insertion (attempting to adjust N_2 will increase the update cost).

4.2.3   The Spatio-Temporal R-Tree

The Spatio-Temporal R-Tree(STR-tree), proposed by Dieter Pfoser et al. in [14], is an extension of R-Tree to support trajectory identity. Generally, STR-tree is a balanced tree with node entries in the format of (ptr, MBR), which is similar to R-Tree. However, as R-Tree cannot be used for indexing 3-dimensional trajectory data, STR-tree adopts a totally different insertion and split mechanism to achieve trajectory orientation.

The insertion process in the STR-tree considers not only *spatial closeness* but also partial *trajectory preservation*, i.e., to keep line segments belonging to the same trajectory together in the index entries. In order to ensure trajectory preservation, STR-tree involves a new algorithm, **FindNode**, to find the proper node that contains the predecessor. Figure 4.4 shows an example of insertion on the STR-tree, where a leaf node returned by **FindNode** is marked with an arrow and p is a preservation parameter to indicate the number of levels for the preservation of trajectories.

The insertion process works as follows: Firstly, a leaf node is returned by **FindNode**, and the new segment is inserted there if there is room in this node. Otherwise, as a node must be split, the insert algorithm checks whether the p-1 parent nodes are full (in Fig. 4.4, for $p = 2$, only the node at non-leaf level 1 needs to be checked). In case one of them is not full, the leaf node is split. In case that all of the p-1 parent nodes are full, the STR-tree would choose another leaf node on

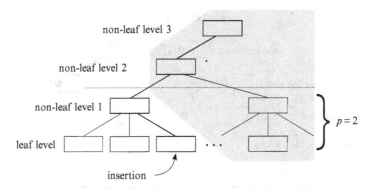

non-leaf level 3

non-leaf level 2

non-leaf level 1

$p = 2$

leaf level

insertion

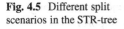

Fig. 4.4 Insertion of STR-tree

Fig. 4.5 Different split
scenarios in the STR-tree

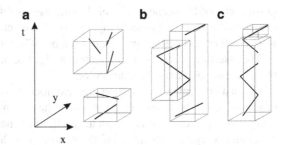

the subtree, including all the nodes further to the right of the current insertion path
(the gray-shaded tree in Fig. 4.4), and then insert as R-Tree.

The general idea of split in the STR-tree is to put newer and thus more recent
segments into new nodes. Splitting non-leaf nodes is simple because it only needs
to create a new node for a new entry. Splitting leaf nodes is shown in Fig. 4.5: If leaf
node contains all disconnected segments, then split as Fig. 4.5a; if leaf node contains
disconnected and other types of segments, then put all disconnected segments in a
new node and split as Fig. 4.5b; otherwise, if it contains single and disconnected
segments, then put the newest single connected segment in a new node and split as
Fig. 4.5c.

Using this insertion and split strategy, STR-tree is an index that preserves
trajectories and considers time as the dominant dimension when decomposing the
occupied space.

4.2.4   The Trajectory-Bundle Tree

The Trajectory-Bundle tree(briefly TB-tree), proposed by Dieter Pfoser et al. in [14],
is an index method that strictly preserves trajectories such that a leaf node only
contains segments belonging to a same trajectory.

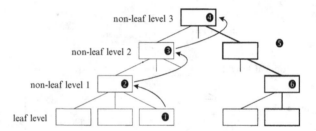

Fig. 4.6 Insertion into the TB-tree

Indexing on trajectory data has to find a balance between spatial discrimination and trajectory preservation. The TB-tree gives up the space discrimination to guarantee trajectory preservations, having spatially close line segments to be stored in different nodes because of their belongings to different trajectories. Though it increases the cost of classical range query, this mechanism is important because it works well on answering "pure spatio-temporal" queries.

The insertion procedure is illustrated in Fig. 4.6. Important stages throughout the procedure are marked with black color.

Given a new entry for insertion, the leaf node that contains its predecessor in the trajectory has to be found. The tree is searched from the root and stepped down to every child node that overlaps with the MBR of the new line segment. The leaf node containing a segment connected to the new entry is chosen (stage 1 in Fig. 4.6). The finding of a segment is similar to that of the STR-tree. If the leaf node is full, a node split is needed. However, due to the violation to trajectory preservation, a new node is actually created in STR-tree instead. Thus, as seen in Fig. 4.6, the tree is searched upwards until a non-full parent node is found (stages 2–4). The rightmost path (stage 5) is chosen to insert the new node. If there is room in the parent node (stage 6), then insert the new leaf node as Fig. 4.6. Otherwise, the node would be split by creating a new node at non-leaf level 1 and having the new leaf node as its only descendant. The split is propagated upwards when necessary. Therefore, the TB-tree grows from left to right; the leftmost leaf node was the first inserted node and the rightmost was the last inserted node.

4.2.5 The MON-Tree

The MON-tree, proposed by Victor Teixeira de Almeida and Güting in [2], is an index structure to store and retrieve past trajectory of network-constrained moving objects. It assumes that moving objects move along polylines, which belong to edges or routes. The MON-tree is composed by a 2D R-Tree (the top R-Tree) indexing polylines bounding boxes and a set of 2D R-Trees (the bottom R-Trees) indexing the movement of objects along the polylines.

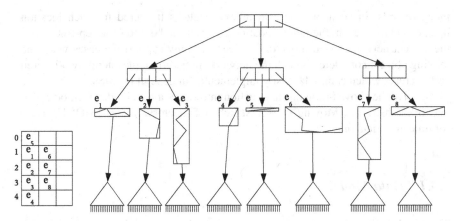

Fig. 4.7 MON-tree index structure

Entries in the top level are of the form $< polyid, bottreept >$, where *polyid* is the polyline identification and *bottreept* is a pointer to the corresponding bottom R-Tree. Moreover, a hash structure organized by *polyid* is used in the top level. Thus, MON-tree is a two-level index structures the R-Tree and a hash structure pointing to bottom level R-Trees. The reason for its two-leveled structures is that the insertion algorithm of moving objects takes a polyline identification as an argument and then uses the top level hash structure to find the place of bottom level R-Tree for insertion. On the other hand, the search algorithm takes a spatio-temporal window as an argument and starts the search on the top R-Tree, which contains the polylines' bounding boxes.

Figure 4.7 is an example of the MON-tree index structure. In the top R-Tree, the polylines are indexed using an MBR approximation. In this way, the leaf nodes contain entries of the form $< mbr, polypt, treepti >$, where *mbr* is the MBR of the polyline, *polypt* points to the real representation of the polyline, and *treept* points to the corresponding bottom R-Tree of that polyline. Internal nodes contain entries of the form $< mbr, childpt >$, where *mbr* is the MBR that contains all MBRs of the entries in the child node and *childpt* is a pointer to the child node.

The bottom R-Tree indexes the movement of moving objects inside a polyline. The movement is represented by the position interval (p_1, p_2) and a time interval (t_1, t_2), where $0 \le p_1, p_2 \le 1$. These two values p_1 and p_2 store the relative position of the objects inside the polyline at times t_1 and t_2, respectively.

4.3   Network-Constrained Moving Object Sketched-Trajectory R-Tree

The moving object index problem has been intensely studied in recent years with a lot of methods proposed, especially the network-based trajectory index methods. However, most existing trajectory index methods take trajectory units as the basic index records. In such methods, an index update is triggered for each location update, and as a result, the index updating cost is turned out to be acceptable. Also, the current network-based trajectory indices can only support the cases when the moving objects completely match the network. In addition, they adopt two-layered architectures which cannot be easily implemented in general DBMSs.

To solve the above problems, this section introduces a novel index method called network-constrained Moving Object Sketched-Trajectory R-Tree (MOSTR-Tree) for trajectory data indexing.

4.3.1 Data Model

The MOSTR-Tree index structure is based on such a data model: A traffic network N is defined as a set of routes R and a set of junctions J, and network position is defined as $npos$, which is the same as defined in chapter Moving Objects Modeling.

Definition 4.1. A *motion vector mv* is a snapshot of moving objects movement at a certain time instant and it is defined as follows:

$$mv = (t, (x, y), v, d, npos)$$

where t is a time instant; $(x, y), v, d$ are the location, the speed, and the direction of the moving object at time t, respectively; and $npos$ is the network position of the moving object at time t.

If $npos \neq \perp$ ($\perp$ means "undefined"), mv is called "network matched." If $npos = \perp$, then mv is not network matched.

Definition 4.2. The *trajectory of a moving object, traj*, is a sequence of motion vectors sent by the moving object through location updates during its journey and is defined as follows:

$$traj = (mv_i)_{i=1}^n = ((t_i, (x_i, y_i), v_i, d_i, npos_i))_{i=1}^n$$

Two neighboring motion vectors of the trajectory, mv_i and mv_{i+1} ($1 \leq i \leq n-1$), can form a trajectory unit, denoted as $\mu(mv_i, mv_{i+1})$. Depending on whether mv_i and mv_{i+1} are network matched, $\mu(mv_i, mv_{i+1})$ can correspond to different shapes in the $X \times Y \times T$ space. If mv_i and mv_{i+1} are both network matched, then $\mu(mv_i, mv_{i+1})$ describes the movement from mv_i to mv_{i+1} along the shortest path between $npos_i$ and $npos_{i+1}$, i.e., the curve in the $X \times Y \times T$ space (see $\mu(mv_1, mv_2)$ and $\mu(mv_4, mv_5)$ in Fig. 4.8). If one of or both mv_i and mv_{i+1} are not network matched, then $\mu(mv_i, mv_{i+1})$ corresponds to a straight line segment in the $X \times Y \times T$ space (see $\mu(mv_2, mv_3)$ and $\mu(mv_3, mv_4)$ in Fig. 4.8).

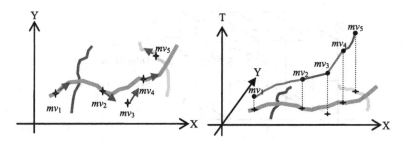

Fig. 4.8 Movement of a moving object and the corresponding trajectory

4.3.2 Index Structure

To transform trajectories to sketched ones, the $X \times Y \times T$ space should be partitioned into several grid cells first. Assume the grid cells are equal sized ones, each of them can be identified by a triple (N_x, N_y, N_t), where N_x, N_y, and N_t are the cell's corresponding serial numbers along the X, Y, T axles. For instance, the gray-colored grid cell in Fig. 4.9 is identified as $(4, 3, 2)$.

Based on the space partition, each trajectory can be transformed into its corresponding sketched trajectory. Here, the trajectory discussed in Definition 4.2 is called "original trajectory" for convenience.

Definition 4.3. Suppose that the original trajectory of a moving object is $traj = ((t_i, (x_i, y_i), v_i, d_i, npos_i))_{i=1}^n$. $traj$'s sketched trajectory, denoted as $sketch(traj)$, is defined as follows:

$$sketch(traj) = (c_j)_{i=1}^n = ((t_j, x_j, y_j))_{i=1}^n$$

where $c_j = (t_j, x_j, y_j)(1 \le j \le k)$ is the center's coordinate of the jth grid cell that $traj$ travels through. Two neighboring coordinates c_j and $c_{j+1}(1 \le j \le k-1)$ of $sketch(traj)$ form a Sketched-Trajectory Unit (STU), denoted as (c_j, c_{j+1}), which corresponds to a straight line segment connecting c_j and c_{j+1} in the $X \times Y \times T$ space. A sketched trajectory can be seen as a sequence of sketched trajectory units so that it forms a polyline in the $X \times Y \times T$ space, as shown in Fig. 4.9.

As depicted in Fig. 4.9, the sketched trajectory approximates to the shape of the original trajectory but has much less trajectory units. Given an original trajectory $traj$, it is obvious that the number of the sketched trajectory units in $sketch(traj)$ is in reverse proportion to the size of grid cells.

Algorithm 4 describes the procedure of transforming an original trajectory to a corresponding sketched trajectory. In this algorithm, function $getCellLocated\ (mv)$ returns the grid cell that mv relates to; Function $getCellsTravelled(\mu)$ returns the grid cell sequence trajectory unit μ travels through; Function $extractCell(cellseq, i)$ extracts the ith grid cell from a grid cell sequence $cellseq$; Function $getCenter(cell)$ returns the center's coordinate of a grid cell cell; $|cellseq|$ returns the number of cells

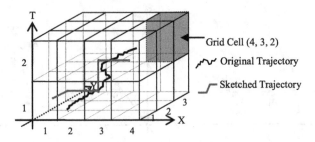

Fig. 4.9 Partition of grid cells and the resulted sketched trajectory

in a grid cell sequence *cellseq*; and function *doNothing()* simply returns without doing anything.

In Algorithm 4, the trajectory units of *traj* are processed one by one. In dealing with a new trajectory unit $\mu(mv_{i-1}, mv_i)$, the algorithm first computes the grids *cell(s)* that the unit travels through by calling *getCellsTravelled()* function and then appends the grid cell *center(s)* to *sketchTraj*.

Algorithm 4: Transforming original trajectory into sketched trajectory

---

input : Spatio-temporal Range of the database: $I_x \times I_y \times I_t$;
Parameters describing the size of grid cells: ξ_x, ξ_y, ξ_t ;
$traj = (mv_i)_{i=1}^n = ((t_i, (x_i, y_i), v_i, d_i, npos_i))_{i=1}^n$
output: $sketchTraj = ((t_j, x_j, y_j))_{j=1}^k$
sketchTraj=NULL;
startingCell = *getCellLocated*(mv_1);
append(*sketchTraj*, *getCenter*(*startingCell*));
if $n = 1$ **then**
  | **return** *sketchTraj*;
else
  currentCell = *startingCell*;
  for $i = 2$ *to* n **do**
    cellsTravelled = *getcellsTravelled*($\mu(mv_{i-1}, mv_i)$);
    if $(|cellsTravelled| = 1)AND(extractCell(cellsTravelled, 1) = currentCell)$ **then**
      | *doNothing*();
    else
      for $j = 2$ *to* $|cellsTravelled|$ **do**
        | *apppend*(*sketchTraj*, *getCenter*(*extractCell*(*cellsTravelled*, *j*)));
      end
      currentCell = *extactCell*(*cellsTravelled*, |*cellsTravelled*|);
    end
  end
  return *sketchTraj*;
end

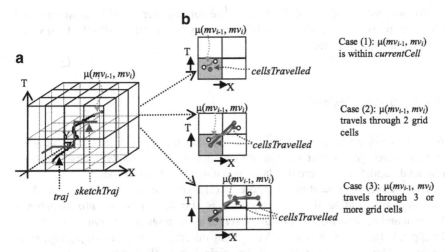

Fig. 4.10 Transforming original trajectory to sketched trajectory. (**a**) Original and Sketched trajectories in $X \times Y \times T$ space. (**b**) Three cases in dealing with $\mu(mv_{i-1}, mv_i)$ ($X \times T$ plane)

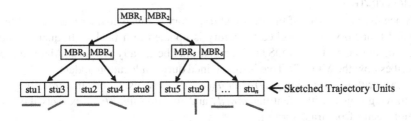

Fig. 4.11 Structure of the MOSTR-Tree

Figure 4.10 shows three typical cases in dealing with a new trajectory unit $\mu(mv_{i-1}, mv_i)$. The gray-colored grid cells in Fig. 4.10 are the cells in which mv_{i-1} is located (i.e., *currentCell*); in case (1), $\mu(mv_{i-1}, mv_i)$ is still inside *currentCell* and nothing will be done in this case. In case (2) and case (3), $\mu(mv_{i-1}, mv_i)$ travels through 2 or more grid cells, and therefore, the center's coordinates of the cells (except *currentCell*) are appended to *sketchTraj*.

After the original trajectories are transformed to sketched ones, the sketched trajectory units can be organized into an R-Tree so that the MOSTR-Tree can be constructed. Figure 4.11 depicts the structure of the MOSTR-Tree.

The leaf nodes of MOSTR-Tree are in the form of $< stu, MBR, PT_{mo} >$, where *stu* is a sketched trajectory unit, *MBR* is the MBR of *stu*, and PT_{mo} is the pointer or identifier leading to the complete database record of the corresponding moving object. Internal nodes of the MOSTR-Tree contain records in the form of $< MBR, PT_{node} >$, where *MBR* is the MBR bounding all the MBRs of the records in its child node and PT_{node} is a pointer leading to the child node.

When constructing the MOSTR-Tree, the database server will transform every trajectory to its corresponding sketched trajectory and insert the sketched trajectory units to the MOSTR-Tree.

4.3.3 Index Update

As moving objects frequently send location update messages to the server, the scale of trajectories grows over time at the server side. Accordingly, index has to be efficiently updated to ensure that latest updating record can be found.

Given a moving object mo, suppose its original trajectory is $traj = (mv_i)_{i=1}^{n}$ and its corresponding sketched trajectory is $sketch(traj)$. When mo launches a location update, it sends to the database server a new motion vector mv_u, which will be appended to $traj$ by the server. When doing this, the server checks if $\mu(mv_n, mv_u)$ has traveled across the boundary of the original grid cell. If not, the appending of mv_u to $traj$ does not trigger the $sketch(traj)$ (MOSTR-Tree) to update. Otherwise, the new sketched trajectory unit(s) corresponding to $\mu(mv_n, mv_u)$ need to be inserted to MOSTR-Tree.

Since the granularity of the sketched trajectory is much coarser than that of the original trajectory, the sketched trajectory is updated in a far less frequency, and the updating cost of the MOSTR-Tree can thus be greatly reduced. Algorithm 5 describes how the MOSTR-Tree is maintained during a location update.

In Algorithm 5, function $getTrajectory(moid)$ retrieves the original trajectory of the moving object with identifier $moid$, and function $final(traj)$ extracts the last motion vector from trajectory $traj$.

Algorithm 5: Maintaining the MOSTR-Tree when receiving a location update message

---

input : Location Update Message: $LUMsg = (moid, t, x, y, v, d, npos)$, MOSTR-Tree: $mostrTree$

$mv_u = (t, (x, y), v, d, npos)$;
$mv_n = final(getTrajectory(moid))$;
$currentCell = getCellLocated(mv_n)$;
$cellsTravelled = getCellsTravelled(\mu(mv_n, mv_u))$;
if $(|cellsTravelled| = 1)AND(extractCell(cellsTravelled, 1) = currentCell)$ **then**
  | $doNothing()$;
else
  | **for** $j = 2$ *to* $|cellsTravelled|$ **do**
  | | $skecthUnit = (getCenter(currentCell), getCenter(extractCell(cellsTravelled, j)))$;
  | | $insert(mostrTree, skecthUnit)$;
  | | $currentCell = extactCell(cellsTravelled, j)$;
  | **end**
end

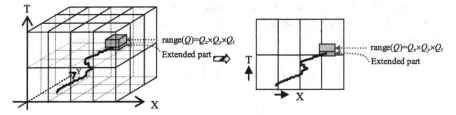

Fig. 4.12 Structure of the MOSTR-Tree

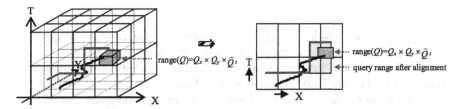

Fig. 4.13 Alignment of the query range to grid cell centers

When constructing and maintaining the MOSTR-Tree with ongoing updates, these location update messages are temporarily saved in buffer. After the MOSTR-Tree is constructed, all the buffered location update messages are then processed with through Algorithm 5 until the buffer is empty, and then the server accepts new location update messages directly and maintain the MOSTR-Tree accordingly.

4.3.4 Query

This part describes the query processing mechanism of the MOSTR-Tree. Suppose that Q is an arbitrary query on moving object trajectories with a query range $(Q) = Q_x \times Q_y \times Q_t$, where $Q_x = [q_x^0, q_x^1]$, $Q_y = [q_y^0, q_y^1]$, and $Q_t = [q_t^0, q_t^1]$. Query range describes X, Y, T ranges that the query concerns and corresponds to a cube in the $X \times Y \times T$ space.

In dealing with such a query, to guarantee all valid objects can be returned, the query time range Q_t should be extended to $\hat{Q}_t = [q_t^0 - \tau, q_t^1]$, as it can be imagined that the original trajectory has a vertical line segment of length τ following the last trajectory unit, where this vertical line segment is missing (not expressed as a record) in the index. Figure 4.12 shows the extension of query range.

After the query range is extended, the system needs to align the query range to grid cell centers, as shown in Fig. 4.13.

In Fig. 4.13, the original trajectory of the moving object intersects range $(Q) = Q_x \times Q_y \times \hat{Q}_t$ and should be included in the query result. However, the sketched trajectory does not intersect with range (Q). To solve this problem, the query range should be adjusted. From the analysis, it can be seen that if the query range is aligned to the corresponding grid cell centers, then all moving objects whose original trajectories intersect the query range, their sketched trajectories will intersect the aligned query range.

Let us first consider $Q_x = [q_x^0, q_x^1]$. Q_x can be transformed as the following:

$$q_x^0 = \left(\left\lfloor \frac{q_x^0 - x_0}{\xi_x} \right\rfloor + 0.5 \right) \times \xi_x$$

$$q_x^1 = \left(\left\lfloor \frac{q_x^1 - x_0}{\xi_x} \right\rfloor + 0.5 \right) \times \xi_x$$

$Q_y = [q_y^0, q_y^1]$ and $\hat{Q}_t = [q_t^0 - \tau, q_t^1]$ can make similar transformations, so that the query range is aligned to the corresponding grid cell centers.

After the alignment transformation is conducted, the whole query processing procedure based on the MOSTR-Tree is provided in Algorithm 6.

In Algorithm 6, the function $timeExend(Q_t, \tau)$ extends the query time range for τ time as described above, function $gridCenterAlign(Q_x \times Q_y \times \hat{Q}_t)$ aligns the query range to grid cell centers and returns the aligned query range as result, function $search(mostrTree, R)$ search the MOSTR-Tree $mostrTree$ according to the specified query range R, function $getTuple(PT_{mo})$ retrieves the tuple from the moving objects database according to PT_{mo}, and function $evaluate(moTuple, Q)$ evaluates the query Q based on the moving object tuple $moTuple$. The computational result is returned if moTuple satisfies the query condition. Otherwise, it simply returns NULL.

Algorithm 6: Query processing based on MOSTR-Tree

---

   input : the Query whose Query Range is range$(Q) = Q_x \times Q_y \times Q_t$: Q, $mostrTree$, Time
            Interval: τ
   output: Query Result: *refineResult*
   $\hat{Q}_t = timeExend(Q_t, \tau)$;
   filterResultsearch$(mostrTree, gridCenterAlign(Q_x \times Q_y \times \hat{Q}_t))$;
   refineResult $= \varnothing$;
   for $\forall PT_{mo} \in filterResult$ **do**
      | $moTuple = getTuple(PT_{mo})$;
      | **if** $evaluate(moTuple, Q) \neq NULL$ **then**
      | | *refineResult* $= refineResult \cup evaluate(moTuple, Q)$;
      | **end**
   end
   return *refineResult*;

As described in Algorithm 6, the query processing based on the MOSTR-Tree consists of two phases: the filtering phase (lines 1 and 2) and the refinement phase (lines 3–10). In the filtering phase, a set of moving objects are retrieved from the

MOSTR-Tree according to the query range, and the results are kept at *filterResult*. In the refinement phase, the moving objects contained in *filterResult* are further evaluated according to the query, and the computation results are kept in *refineResult* which is finally returned to the querying user.

According to the Algorithm 6, the overall query processing time depends on both the index retrieval time and tuple refinement time. The index retrieval time is affected by the index updating cost and the index searching cost. From analysis it can be seen that the larger the grid cell size is, the lower the index updating cost will be, and the more the useless records will be contained in *filterResult*, leading to greater cost in tuple refinement, and vice versa. Thus, to get the best overall query processing performance, a suitable grid cell size should be chosen.

4.4   Network-Constrained Moving Objects Dynamic Trajectory R-Tree

In this section, we would introduce a new index structure for network-constrained moving objects, Network-constrained moving objects Dynamic Trajectory R-Tree (NDTR-Tree), which can deal with not only the historical locations of moving objects but also their current and near-future location information. The NDTR-Tree employs a hybrid structure with two layers. Its upper layer is edge-based R-Tree, which indexes the directed atomic route sections with smaller granularity, so that the intersection between different MBRs can be greatly reduced, while its lower layered R-Trees are route based, with each lower R-Tree corresponding to a route which has a greater granularity, so that location update and index maintaining costs can be reduced. In this way, the query processing and index maintaining performances can be improved.

For better understanding the NDTR-tree, a two-layered, route-ARS-based traffic network framework is introduced first, which compromises the network constrained moving objects and trajectories.

A traffic network N is defined as a set of routes R and a set of junctions J, a directed atomic route section (ARS) is defined as ars, and a network position is defined as $npos$. Based on the traffic network, a motion vector mv is a snapshot of moving object's movements, and the trajectory of moving object is a sequence of motion vectors, which is denoted as Tr.

4.4.1   Index Structure of NDTR-Tree

The NDTR-Tree is two-layered structure. The upper layer is a single R-Tree which indexes the directed atomic route sections of the traffic network, and the lower layer consists of a forest of R-Trees, with each R-Tree corresponding to a certain route

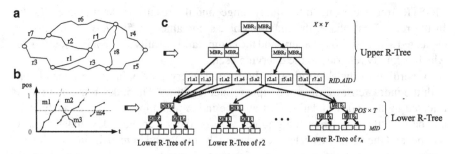

Fig. 4.14 Structure of the NDTR-Tree. (**a**) Traffic network. (**b**) UT-Units submitted in route $r1$. (**c**) The corresponding NDTR-Tree

and indexing the trajectory units submitted in the route. Figure 4.14 illustrates the structure of the NDTR-Tree.

As shown in Fig. 4.14, the upper layer of the NDTR-Tree is a standard R-Tree which takes directed atomic route sections as the basic structural unit. The records of the leaf nodes take the form $< MBR_{xy}, rid.aid, pt_{route}, pt_{tree} >$, where MBR_{xy} is the two-dimensional MBR that covers directed atomic route sections, $rid.aid$ is a combination of the route and ARS identifiers, pt_{route} is a pointer to the detailed route record, and pt_{tree} is a pointer to the lower R-Tree corresponding to $route(rid)$. The root or internal nodes contain records of the form $< MBR_{xy}, pt_{node} >$, where MBR_{xy} is the MBR (in the $X \times Y$ plane) containing all MBRs of its child records and pt_{node} is a pointer to the child node.

The lower layer of the NDTR-Tree is composed of a set of R-Trees. Each R-Tree only corresponds to a route (e.g., John Street) and indexes all the trajectory unit on it. The format of the leaf nodes is $< MBR_{pt}, mid, mv_s, mv_e >$, where MBR_{pt} is the MBR of the associated trajectory units, mid is the identifier of the moving object, and $mv_s = (t_s, rid, pos_s, \overline{v}_s)$ and $mv_e = (t_e, rid, pos_e, \overline{v}_e)$ are the two consecutive motion vectors which form the trajectory units. If the trajectory is active unit, then mv_e is NULL and the MBR is $< t_a, pos_a, min(t_\xi, t_\psi), 1 >$. The format of both root and internal nodes is $< MBR_{pt}, pt_{node} >$, where MBR_{pt} is the MBR covering all MBRs of its child records and pt_{node} is the pointer to its child node.

4.4.2 Active Trajectory Unit Management

To support queries on current and future position of moving objects, it is necessary to put active trajectory unit in the NDTR-Tree. However, as an active trajectory unit $\mu(mv_n)$ is essentially a ray l between points (pos_n, t_n), and the slope of l is determined by $\overrightarrow{v_n}$. Thus, the first problem is setting of length of l. In the following discussion, we assume an object mo moves from the starting node 0 towards the end node 1 of a route r. Let point $(t*, 1)$ to be the intersection between l and $pos = 1$; $t*$ can be easily computed as

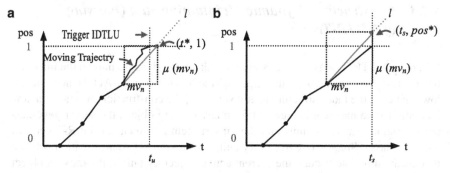

Fig. 4.15 Determining the MBRs of active trajectory units. (**a**) Speed $> v_n$. (**b**) Speed $< v_n$

$$t* = t_n + \frac{(1 - pos_n) \times r.length}{|\vec{v_n}|}$$

where *r.length* is the length of route r. If *mo* does not update its location before reaching the end node of r (i.e., to move as the predicted trajectory), there are generally two possible cases: (1) its actual speed $\vec{v} \geq \vec{v_n}$. According to Fig. 4.15a, an IDTLU update is triggered at the time t_u in such case, therefore the computing of *mo*'s current position only relevant to the section of ray l under $pos = 1$ (the full line). (2) its actual speed $\vec{v} < \vec{v_n}$. Based on the updating policy, the minimum speed of *mo* is $\vec{v_s} = \vec{v_n} - \vec{v_t}$ ($\vec{v_t}$ is a speed threshold). Based on the time $\vec{v_s}$, the time t_s needed for reaching end node 1 can be calculated as:

$$t_s = t_n + \frac{(1 - pos_n) \times r.length}{|\vec{v_s}|}$$

The intersection between ray l and line $t = t_s$ is (t_s, pos_s), and pos_s can be calculated as

$$pos_s = pos_n + \frac{(t_s - t_n) \times |\vec{v_n}|}{r.length}$$

As shown the full line of Fig. 4.15b, current location of *mo* in the second case could be at any position on l below the point (t_s, pos_s).

Based on (1) and (2), the NDTR-Tree index only need to preserve the active trajectory unit as the $MBR = < t_n, pos_n, t_s, pos_s >$ corresponding to the ray l, and this region is sufficient to answer both current and future position queries over moving objects, where

$$t_s = t_n + \frac{(1 - pos_n) \times r.length}{\vec{v_s}}, \quad pos_s = pos_n + \frac{(t_s - t_n) \times |\vec{v_n}|}{r.length}$$

4.4.3 Constructing, Dynamic Maintaining, and Querying of NDTR-Tree

When the NDTR-Tree is first constructed in database, the system read the information of road network and build the upper R-Tree based on routes. At this moment, the lower layer of the index structure is null yet. After the construction, whenever a new location update message is received from any moving object, the server generates corresponding trajectory units and then insert them into their related R-Tree(s) in the lower layer. Since active trajectory units contain predictive information, when a new location update occurs, the current active trajectory unit of the moving object should be replaced by newly generated records.

The construction and dynamic maintenance algorithm for the NDTR-Tree is given in Algorithm 7, where function $MBR(u)$ returns the MBR of a given trajectory unit u and functions $Insert()$ and $Delete()$ means to conduct the insertion and deletion of trajectory units in the corresponding lower R-Trees, respectively.

Algorithm 7: Constructiion and dynamic maintenance algorithm of NDTR-Tree

---

input : Traffic Network: $N = (Routes, Junctions)$
Read route records of N, and insert the related ARSs into the upper R-Tree;
Set all lower R-Trees to empty tree;
while *MOD is running* **do**

    Receive location update package LUM from moving objects (Suppose the moving object ID is mid);

    if *LUM contains 1 motion vector $mv_a = (t_a, rid_a, pos_a, \overline{v}_a)$* **then**

        Let mv_n be the current active motion vector of mo in $RTree_{low}(rid_a)$;

        if $mv_n = NULL$ **then**

            Insert($RTree_{low}(rid_a), (mid, \mu(mv_n), \text{mbr}(\mu(mv_n))))$;

        else

            Delete($RTree_{low}(rid_a), (mid, \mu(mv_n), \text{mbr}(\mu(mv_n))))$;

            Insert($RTree_{low}(rid_a), (mid, \mu(mv_n, mv_a), \text{mbr}(\mu(mv_n, mv_a))))$;

            Insert($RTree_{low}(rid_a), (mid, \mu(mv_a), \text{mbr}(\mu(mv_a))))$;

        end

    else if *LUM contains 3 motion vectors $mv_{a1}, mv_{a2}, mv_{a3}$* **then**

        Let mv_n be the current active motion vector of mo in $RTree_{low}(rid_{a1})$;

        Delete($RTree_{low}(rid_{a1}), (mid, \mu(mv_n), \text{mbr}(\mu(mv_n))))$;

        Insert($RTree_{low}(rid_{a1}), (mid, \mu(mv_n, mv_{a1}), \text{mbr}(\mu(mv_n, mv_{a1}))))$;

        Insert($RTree_{low}(rid_{a2}), (mid, \mu(mv_{a2}, mv_{a3}), \text{mbr}(\mu(mv_{a2}, mv_{a3}))))$;

        Insert($RTree_{low}(rid_{a2}), (mid, \mu(mv_{a3}), \text{mbr}(\mu(mv_{a3}))))$;

    end

end

Since in moving objects databases, the most common query operators, such as possibly inside (trajectory, $I_x \times I_y \times I_t$) and possibly intersect (trajectory, $I_x \times I_y \times I_t$) (where I_x, I_y, I_t are intervals in X, Y, T domains), belong to range queries, that is,

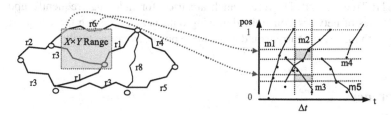

Fig. 4.16 Range query through NDTR-Tree. (**a**) Search the upper R-Tree and receive (*rid×period*) pairs. (**b**) Search the lower R-Trees and output moving object identifiers (*m2, m3*)

the input of the query is a range in the $X \times Y \times T$ space, we take range query as an example to show how the query processing is supported by the NDTR-Tree.

We use the most common query type, i.e., spatial range queries over moving objects, as example to illustrate and analyze the query processing. The querying of the NDTR-Tree can be finished in two steps. When processing a range query (suppose the range is $I_x \times I_y \times I_t$), the system will first query the upper R-Tree of the NDTR-Tree according to $I_x \times I_y$ and will receive a set of *(rid, period)* pairs as the result, where *period* $\in [0, 1]$ and can have multiple elements; then for each *(rid, period)* pair, search the corresponding lower R-Tree to find the trajectory units intersecting *period* $\times I_t$ and output the corresponding moving object identifiers. Figure 4.16 illustrates the range query processing based on the NDTR-Tree.

The query algorithm is given in Algorithm 8.

Algorithm 8: Range query algorithm of NDTR-Tree

---

input : Quering Range: $I_x \times I_y \times I_t$
output: Set of moving objects identifiers: *Result*
Search the upper R-Tree according to $I_x \times I_y$, and receive a set of pairs: $(rid_i, period_i)_{i=1}^{n}$;
for $1 \leq i \leq n$ **do**
    for $\forall \rho \in period_i \times I_t$ **do**
        Let μ be the set of trajectory units in $RTree_{low}(rid_i)$ which intersect ρ;
        Result= Result$\cup$ the set of moving object IDs contained in the element of μ;
    end
end
return *Result*;

4.5 Summary

In this chapter, we discuss the indexing for moving objects; several representative indexing methods are introduced for moving objects including R-Tree, TPR-tree, STR-tree, TB-tree, and MON-tree. We proposed two new index methods,

MOSTR-Tree and NDTR-Tree, which are used for indexing frequently updated trajectories of network-constrained moving objects and indexing the whole trajectories with historical, current, and near-future positions, respectively.

References

1. Agarwal PK, Arge L, Erickson J (2000) Indexing moving points. In: Proceedings of the 19th ACM SIGMOD-SIGACT-SIGART symposium on principles of database systems (PODS 2000), Dallas, pp 175–186
2. Almeida VT, Güting RH (2005) Indexing the trajectories of moving objects in networks. GeoInformatica 9(1):33–60
3. Beckmann N, Kriegel HP, Schneider R, Seeger B (1990) The R*-tree: an efficient and robust access method for points and rectangles. In: Proceedings of the 1990 ACM SIGMOD international conference on management of data (SIGMOD 1990), Atlantic City, pp 322–331
4. Chen J, Meng X (2009) Update-efficient indexing of moving objects in road networks. GeoInformatica 13(4):397–424
5. Chen J, Meng X, Guo Y, Grumbach S (2007) Indexing future trajectories of moving objects in a constrained network. J Comput Sci Technol 22(2):245–251
6. Ding Z (2011) Indexing frequently updated trajectories of network-constrained moving objects. In: Proceedings of DEXA, Toulouse, pp 464–474
7. Ding Z, Li X, Yu B (2009) Indexing the historical, current, and future locations of network-constrained moving objects. J Softw 20(12):3193–3204 (in Chinese)
8. Frentzos E (2003) Indexing objects moving on fixed networks. In: Proceedings of the 8th international symposium on spatial and temporal databases (SSTD 2003), Santorini Island, pp 289–305
9. Guttman A (1984) R-trees: a dynamic index structure for spatial searching. In: Proceedings of the ACM SIGMOD international conference on management of data (SIGMOD 1984), Boston, pp 47–57
10. Jensen CS, Lin D, Ooi BC (2004) Query and update efficient B^+ tree based indexing of moving objects. In: Proceedings of the 30th international conference on very large data bases (VLDB 2004), Toronto, pp 768–779
11. Kollios G, Gunopulos D, Tsotras VJ (1999) Effective density queries on continuously moving objects. In: Proceedings of the 22nd international conference on data engineering (ICDE 1999), Atlanta, p 71
12. Nascimento MA, Silva JRO (1998) Towards historical R-trees. In: ACM symposium on applied computing (SAC 1998), Atlanta, pp 235–240
13. Pfoser D, Jensen CS (2003) Indexing of network constrained moving objects. In: Proceedings of the 11th ACM international symposium on advances in geographic information systems (GIS 2003), New Orleans, pp 25–32
14. Pfoser D, Jensen CS, Theodoridis Y (2000) Novel approaches in query processing for moving object trajectories. In: Proceedings of the 26th international conference on very large data bases (VLDB 2000), Cairo, pp 395–406
15. Saltenis S, Jensen CS (2002) Indexing of moving objects for location-based service. In: Proceedings of the 18th international conference on data engineering (ICDE 2002), San Jose, pp 463–472
16. Saltenis S, Jensen CS, Leutenegger ST, Lopez MA (2000) Indexing the positions of continuously moving objects. In: Proceedings of the ACM SIGMOD international conference on management of data (SIGMOD 2000), Dallas, pp 331–342
17. Tao Y, Faloutsos C, Papadias D, Liu B (2004) Prediction and indexing of moving objects with unknown motion patterns. In: Proceedings of the ACM SIGMOD international conference on management of data (SIGMOD 2004), Paris, pp 611–622

Chapter 5
Moving Objects Basic Querying

Abstract Once we build the model and index for moving objects, we can answer the queries for moving objects. There are many types of queries in moving objects databases such as the nearest neighbor (NN) query, range query, and density query. In this chapter, we will introduce the basic querying types for moving objects according to spatial predicates, temporal predicates, and moving spaces. Though there are many techniques to support moving objects queries, most of the existing studies consider Euclidean spaces, where the distance between two objects is determined solely by their relative position in space. However, in practice, objects can usually move only on a predefined set of trajectories as specified by the underlying network. Hence, we will introduce how to answer range queries and NN queries for moving objects in a spatial network, which is based on the work of Papadias in Papadias et al. (Query processing in spatial network databases. In: Proceedings of the 29th international conference on very large data bases (VLDB 2003), Berlin, pp 790–801, 2003).

Keywords Spatio-temporal query • Nearest neighbor query • Range query • Spatial network • Moving object databases

5.1 Introduction

Considerable research has been carried out on moving object databases, which has resulted in the development of numerous indexes and query processing techniques. Surprisingly, most of the existing studies consider Euclidean spaces, where the distance between two objects is determined solely by their relative position in space. However, in many applications that manage spatial data (e.g., location-based services), the position and accessibility of spatial objects are constrained by spatial networks such as road, railway, and river. In such cases, the actual distance between two objects corresponds to the length of the shortest path connecting them in the network, i.e., the network distance.

X. Meng et al., *Moving Objects Management: Models, Techniques and Applications*, DOI 10.1007/978-3-642-38276-5__5,
© Tsinghua University Press, Beijing and Springer-Verlag Berlin Heidelberg 2014

Fig. 5.1 Road network query
example

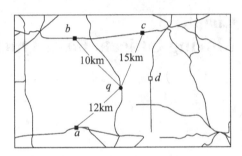

For instance, consider the spatial network of Fig. 5.1, where the rectangles correspond to hotels. If a user at location q poses the range query "find the hotels within a 15 km range", the results will contain a, b, and c (the numbers in the figure correspond to network distance). Similarly, a nearest neighbor query will return hotel b. Note that the results of the corresponding conventional queries are different (e.g., the Euclidean nearest neighbor is d, which is actually the farthest hotel in the network). Furthermore, queries may combine both location and network aspects, such as "find the nearest hotel to the south" (e.g., hotel a).

In this chapter, we will introduce the basic querying types for moving objects according to spatial predicates, temporal predicates, and moving spaces. Then we propose how to process range queries and NN queries for moving objects in a spatial network based on the Euclidean restriction and network expansion frameworks. The resulting algorithms expand conventional processing techniques by integrating connectivity and location information for efficient pruning of the search space [11].

5.2   Classifications of Moving Object Queries

The moving object has two kinds of attributes – spatial and temporal. Therefore, to answer the queries for moving objects, the spatial and temporal predicates must be indicated. The answers for these queries are moving objects that satisfy the predicates. Hence, there are many kinds of queries for moving object data according to spatial and temporal predicates. We introduce them in this section.

5.2.1   Based on Spatial Predicates

The spatial predicate indicates a point or range. The queries for moving objects can be divided into four classes as follows:

1. *Range Query*: A range query is to find the objects within some specific area that corresponds to a rectangular window or a circular area around a query point. For example, "find all of the people who walked within one mile of the buildings at

the time." The range queries are the most basic queries and are widely used. Most moving objects indexing methods can support range query processing.

2. *Nearest Neighbor Query*: A nearest neighbor (NN) query is to find the object which is nearest to a query point. The most popular NN query is kNN query, which is to find the k nearest neighbors to a query point. There is another kind of NN query, called reverse nearest neighbor (RNN) query. The RNN query is to find the object whose nearest neighbor is the query point. For example, consider some taxies and some passengers. The passenger wants to know which taxi is closest to him. The taxi wants to find the passenger who has this taxi as a nearest neighbor, so he will be a possible customer. So far, two kinds of approaches have been developed to process an NN query: index traverse and region pre-computation. Most research studies adopt the first approach and use the R-Tree or Quad-tree to index the moving objects. A typical algorithm is the branch-and-bound algorithm proposed by Roussopoulos et al. in [18]. This approach traverses the R-Tree to find the nearest neighbor of the query point in a depth-first manner. For region pre-computation, the Voronoi graph is a typical method to find the result of an NN query [7].

3. *Aggregate Nearest Neighbor Query*: An aggregate nearest neighbor (ANN) query returns the object that minimizes an aggregate distance function with respect to a set of query points. Consider, for example, several users at specific locations (query points) that want to find the restaurant (data point), which leads to the minimum sum of distances that they have to travel in order to meet. ANN queries are a natural way to express requests by groups of mobile users who want to optimize their routes according to an aggregate function applying on the traveling distances. Apart from the meeting-restaurant example, other application instances include (1) establishing a meeting station for members of a new church based on its distances from their homes and (2) selecting the location of a touristic office based on its distances to attractions in a city. Yiu et al. [22] solve the ANN queries for objects in spatial networks. They consider alternative aggregate functions and techniques that utilize Euclidean distance bounds, spatial access methods, and/or network distance materialization structures.

4. *Density Query*: Density query [4, 6, 10] involves finding dense areas with high concentration of moving objects, where the density of moving objects is higher than the given threshold. Hadjieleftheriou et al. [4] first propose the density query for moving objects. They define density region as $density(R, \Delta t) = min\Delta t N / area(R)$, where $min\Delta t N$ is the minimum total number of objects in region R during time interval Δt; $area(R)$ is the area of R. Based on the definition, they introduced two types of density queries: snapshot density queries (SDQ) and period density queries (PDQ). In the case of SDQ, users require information about the dense regions in a specific time, for example, "tell me the region where the total number of cars is more than 100 at 3 pm." In the case of PDQ, users require information about the dense regions within a time period, for instance, "tell me the region where the total number of cars is always more than 100 in 10 minutes." Jensen et al. [6] focus on how to find the dense regions in a specific time. Similar to the work of [4], they also assume there

are a lot of moving objects in a Euclidean space and these objects move in a linear manner. The difference is that they can avoid the answer loss. Both studies assume the objects to be moving in a freestyle and thereby define the density query in Euclidean space. However, efficient dynamic density query in spatial networks is more crucial for many real-life applications.

5.2.2 Based on Temporal Predicates

There are three kinds of different temporal predicates in moving objects queries. Accordingly, the queries can be classified to three classes: historical, current, and future query. There are different indexes supporting the different queries. In the case of range query, historical indexes such as TB-tree [16] can support range queries for historical data; current indexes such as LUR-tree [8], which is based on R-Tree, can support queries for current locations; future indexes can answer future range queries by predicting the location of moving objects for a limited time period and the query result precision is determined by the prediction model. For historical queries and current queries, the processes are relatively simple; but in the case of future queries, the process is more complex because the future location needs to be predicted. There are two typical approaches for future queries: space transformation technology in multidimension space, such as STRIPES [13] and expanding approach such as the TPR-tree, TPR*-tree, and B^x-tree. The transformation approach divides space into non-overlapping parts and transfers the trajectory in $(d - 1)$ dimension into points in 2-dimensional space. The expanding approach can be divided into two forms: query range expanding and MBR expanding, which is more widely used than the transformation approach.

5.2.3 Based on Moving Spaces

Moving objects queries can be divided into queries in Euclidean spaces and in spatial networks. Most of the existing studies focus on query processing in Euclidean spaces. For query processing in spatial networks, the distance metric is different from the Euclidean distance, and so the method used in Euclidean spaces cannot be used in spatial networks. There are two main differences: "nearest" refers not to the nearby location but the smallest network distance; the distance between objects is not determined by locations of objects but the connection of network.

There are three kinds of approaches for query processing in spatial networks: (1) combining the tree traversing with route searching, (2) applying the multi-pass shortest path algorithm to the network distance computations that starts from a single source to all destinations, and (3) transforming the spatial network to hyperspace and using the Euclidean measurement method. The main idea in these approaches is filtering out the unnecessary objects using some space partitioning

methods to reduce shortest path computation and then refine the candidate set by
network distance to get the final results. The disk-based network representation
method [3] can support NN queries, range queries, and closet pair queries by
combining spatial network connections and Euclidean location information. The
ANN queries in spatial networks can be processed by using the boundary properties of Euclidean distances, spatial data access method, and network distance
materialization technology [22]. In [23], the authors use graph theory and query
result materialization technology to reduce the network expansion in the Dijkstra
algorithm and improve the efficiency of processing RNN queries.

5.3 Point Queries

Given a moving object's trajectory, it is fundamentally important to query its
position in both temporal and spatial dimensions in many scenarios such as fleet
management, air traffic control. As introduced in Chap. 2, Wolfson first proposed
a model to query a point's position on a given trajectory. Based on different data
models, researchers introduced several approaches for querying a point on the
trajectory. In general, the point query involves interpolation between sampling
points. As a result, the accuracy is heavily influenced by the representation of
trajectory. The initial research works take the trajectory between sampling points as
linear. However, in many cases, the objects movements are constrained by underling
transportation networks. For example a bus may move along a specific route. So, the
interaction between the moving object and the underling transportation networks
should be taken into count for the query. For better indexing, the temporal aspect
of moving object is related to the transport network on which interpolation is
implemented.

As is known well, a representation of moving-point trajectories is inherently
imprecise: imprecision is introduced by the measurement process used in the sampling of positions and by the sampling approach itself, therefore these uncertainties
and the subsequent imprecisions should be taken into account when process the
data within the trajectory. Pfoser and Jensen [14] consider the inherent uncertainty
and imprecision within the sample data about the positions of the moving-point
objects by quantifying the error and uncertainty of these samples in their modular
presentation. Then the positions in-between the sampled positions of objects are
obtained to help trace the complete trajectory and deal with trajectory queries with
probability. Mokhtar and Su [9] use a vector of uniform stochastic processes to
model the uncertain trajectories of moving objects, so as to increase the accuracy of
the queries.

Although lots of methods in machine learning exist for prediction, to predict
the future locations of the objects effectively and fast can be a challenge. Recent
studies show that using new data structures can be a potential solution for this. Tao
et al. [21] integrate novel insertion and deletion algorithms to propose the TPR*-
tree algorithm, which is based on the mainstream practical method TPR-tree and

can be nearly optimal when calculating the future queries. Pfoser and Jensen [15] convert the three-dimensional (x, y, t) trajectory data into two-dimensional (x, t) by reducing the movements to occur in one spatial dimension for processing the data. Given the right circumstances, their method performs more efficient than using a three-dimensional index method. Song and Roussopoulos [20] transform the history records of the objects into points and then use SEB-tree to store the history records of one zone. Experiments show this new indexing structure can accelerate the interpolation queries compared to other counterparts.

5.4 NN Queries

Given a source point q and an entity dataset S, a kNN query retrieves the k (≥ 1) objects of S closest to q according to the network distance (e.g., "find the hotel within the shortest driving distance"). This section presents two algorithms for nearest neighbor queries, based on the Euclidean restriction and network expansion frameworks. Euclidean restriction takes advantage of the Euclidean lower-bound property to prune the search space. On the other hand, the network expansion framework performs query processing directly on the network [11].

5.4.1 Incremental Euclidean Restriction

The Incremental Euclidean Restriction (IER) algorithm applies the multi-step kNN methodology [2, 19], traditionally used for high-dimensional similarity retrieval. Specifically, assuming that only one NN is required, IER first retrieves the Euclidean nearest neighbor p_{E1} of q, using an incremental kNN algorithm (e.g., [5]) on the entity R-Tree of S. Then, the network distance $d_N(q, p_{E1})$ of p_{E1} is computed. Owing to the Euclidean lower-bound property, objects closer (to q) than p_{E1} in the network should be within Euclidean distance $d_{Emax} = d_N(q, p_{E1})$ from q, i.e., they should lie in the shaded area of Fig. 5.2a. In Fig. 5.2b, the second Euclidean NN p_{E2} is then retrieved (within the d_{Emax} range). Since $d_N(q, p_{E2}) < d_N(q, p_{E1})$, p_{E2} becomes the current NN and d_{Emax} is updated to $d_N(q, p_{E2})$, after which the search region (for potential results) becomes smaller (the shaded area in Fig. 5.2b). Since the next Euclidean NN p_{E3} falls outside the search region, the algorithm terminates with p_{E2} as the final result.

The extension to k nearest neighbors is straightforward. The k Euclidean NNs are first obtained using the entity R-Tree, sorted in ascending order of their network distance to q, and d_{Emax} is set to the distance of the kth point. Similar to the single NN case, the subsequent Euclidean neighbors are retrieved incrementally, while maintaining the k (network) NNs and d_{Emax} (except that d_{Emax} equals the network distance of the kth neighbor), until the Euclidean distance of the next Euclidean NN is larger than d_{Emax}. Algorithm 9 illustrates the pseudo-code of IER.

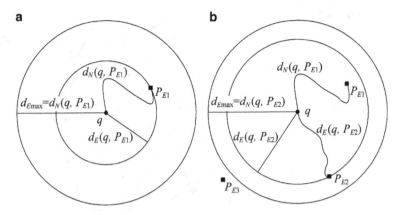

Fig. 5.2 Finding the NN p_{E2}. (**a**) First Euclidean NN. (**b**) Second Euclidean NN

Algorithm 9: IER (q, k)

---

input : q is the query point, k is the number of query results
output: k nearest neighbors to q
$\{p_1, \ldots, p_k\}$=Euclidean_NN(q, k);
for *each entity p_i* **do**
| $d_N(q, p_i) = compute\_ND(q, p_i)$;
end
sort $p_1, \ldots, p_k$ in ascending order of $d_N(q, p_i)$;
$d_{Emax} = d_N(q, p_k)$;
while $d_E(q, p) \leq d_{Emax}$ **do**
| $(p, d_E(q, p))$=next_Euclidean_NN(q);
| **if** $d_N(q, p) < d_N(q, p_k)$ **then**
| | insert p in $\{p_1, \ldots, p_k\}$;
| | $d_{Emax} = d_N(q, p_k)$;
| **end**
end

5.4.2 *Incremental Network Expansion*

IER (and the Euclidean restriction framework in general) is more effective if the ranking of the data points by their Euclidean distance is similar to that with respect to the network distance. Otherwise, a large number of Euclidean NNs may be inspected before the network NN is found. Figure 5.3 shows an example where the black points represent the nodes in the modeling graph and rectangles denote entities. The nearest entity to the query q (white point) is p_5. The subscripts of the entities ($p_1, p_2, \ldots, p_5$) are in ascending order of their Euclidean distance to q. Since p_5 has the largest Euclidean distance, it will be examined after all other entities, i.e., p_1–p_4, correspond to *false hits*, for which the network distance computations are redundant.

Fig. 5.3 Finding the NN p_5

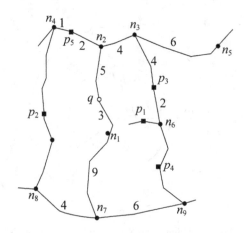

To remedy this problem, the incremental network expansion (INE) algorithm performs network expansion (starting from q) and examines entities in the order they are encountered. Specifically, INE first locates the segment n_1n_2 that covers q and retrieves all entities on n_1n_2. Since no point is covered by n_1n_2, the node (n_1) closest to the query is expanded (while the second endpoint n_2 of n_1n_2 is placed in a queue Q). No data point is found in n_1n_7 and n_7 is inserted to $Q = $ <$(n_2,5)$, $(n_7,12)$>. The expansion of n_2 reaches n_4 and n_3, after which $Q = $ <$(n_4,7)$, $(n_3,9)$, $(n_7,12)$> and point p_5 is discovered on n_2n_4 (while no point is found on n_2n_3). The distance $d_N(q, p_5) = 6$ provides a bound d_{Nmax} to restrict the search space. The algorithm terminates now since the next entry n_4 in Q has larger distance (i.e., 7) than d_{Nmax}. Algorithm 10 shows the pseudo-code of INE.

Algorithm 10: INE (q, k)

---

    input : q is the query point, k is the number of query results
    output: k nearest neighbors to q
    $n_in_j = find\_segment(q)$;
    $S_{cover} = find\_entities(n_in_j)$;
    $\{p_1, \ldots, p_k\}$ = the k (network) nearest entities in S_{cover} sorted in ascending order of their network distance;
    $d_{Nmax} = d_N(q, p_k)$;
    $Q = $ < $(n_i, d_N(q, n_i)), (n_j, d_N(q, n_j))$ >;
    de-queue the node n in Q with the smallest $d_N(q, n)$;
    while $d_N(q, n) < d_{Nmax}$ **do**
        for *each non-visited adjacent node* n_x *of* n **do**
            $S_{cover} = find\_entities(n_xn)$;
            update $\{p_1, \ldots, p_k\}$ from $\{p_1, \ldots, p_k\} \bigcup S_{cover}$;
            $d_{Nmax} = d_N(q, p_k)$;
            en-queue $(n_x, d_N(q, n_x))$;
        end
        de-queue the next node n in Q;
    end

5.5   Range Queries

Given a source point q, a value e, and a spatial dataset S, a range query retrieves all objects of S that are within the network distance e from q. This section applies the Euclidean restriction and network expansion paradigms for processing such queries [11].

5.5.1   Range Euclidean Restriction

The Range Euclidean Restriction (RER) method first performs a range query at the entity dataset and returns the set of objects S' within (Euclidean) distance e from q. Assuming the Euclidean lower-bound property, S' is guaranteed to avoid false misses (i.e., $d_N(q, p) \leq e \Rightarrow d_E(q, p) \leq e$), but it may contain a large number of false hits. In order to reduce the number of network distance computations, RER performs network expansion only once, examining all segments within network distance e from q. Points of S' that fall on some segment are removed from S' and returned to the user. The process terminates when all the segments in the range are exhausted, or when S' becomes empty. Algorithm 11 illustrates the pseudo-code of the algorithm. S' contains the results of the Euclidean range query sorted on some dimension. When a new segment is encountered, the sorted list is used to efficiently check if any point falls inside its MBR (filter step). Such points are then compared with the polyline representation of the segment to determine whether they belong

---

Algorithm 11: RER (q, e)

input : q is the query point, e is the network distance threshold
output: objects of S that are within network distance e from q
$result = \emptyset$;
$S' = $ Euclidean-range(q, e);
$n_i n_j = find\_segment(q)$;
$Q = < (n_i, d_N(q, n_I)), (n_j, d_N(q, n_j)) >$;
de-queue the node n in Q with the smallest $d_N(q, n)$;
while $d_N(q, n) \leq e$ and $S' \neq \emptyset$ **do**
    for *each non-visited adjacent node n_x of n* **do**
        for *each point s of S'* **do**
            if $check\_entity(n_x n, s)$ **then**
                $result = result \bigcup \{s\}$;
                $S' = S' - \{s\}$;
        end
    end
    en-queue$(n_x, d_N(q, n_x))$;
end
de-queue the next node n in Q;
end

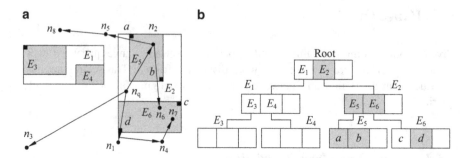

Fig. 5.4 An example of RNE. (**a**) Network and objects. (**b**) The object R-tree

to the actual result (refinement step). Part of some segments at the boundary may exceed the query threshold e, but these segments must be considered nonetheless since they may contain data points that satisfy the query.

5.5.2 Range Network Expansion

The Range Network Expansion (RNE) algorithm first computes the set QS of qualifying segments within network range e from q and then retrieves the data entities falling on these segments. The methodology is similar to INE, but now numerous queries, one for each qualifying segment, are performed simultaneously. To illustrate RNE, assume that QS contains the segments shown in Fig. 5.4a. Starting from the root of the object R-Tree, RNE visits nodes that intersect the MBR of at least one segment in QS. Figure 5.4b illustrates the visited nodes and the qualifying objects in gray.

In order to avoid joining the entire QS (which may be large) with every entry, we perform the following optimization. QS is divided into (possibly overlapping) sets QS_i, one for each entry E_i in the current R-Tree node. A segment is assigned to all entries that intersect its MBR. When the children of E_i are visited, they are only compared against QS_i. Thus, as RNE descends the tree, the number of comparisons performed for each entry is reduced. In Fig. 5.4, the set of qualifying segments $QS_1 = \emptyset$, while for E_2, QS_2 consists of all segments except n_1n_4 and n_5n_8. Similarly, $QS_5 = \{n_qn_2, n_2n_5, n_2n_6\}$ and $QS_6 = \{n_qn_1, n_2n_6, n_4n_7\}$. When the node of $E_5(E_6)$ is visited, its points will only be checked against the segments of $QS_5(QS_6)$.

An object can be reported more than once if it lies at the intersections of the segments in QS. Such duplicates are easy to remove, by sorting the results at each leaf node before reporting them. RNE is I/O optimal (since it only accesses R-Tree nodes that overlap some qualifying segment, and therefore, may contain results).

The pseudo-code of RNE is presented in Algorithm 12. The initial parameters of the algorithm are root of R-tree S, QS, $\emptyset$. To reduce the number of intersection tests, at lines 2 and 7, we apply a plane sweep algorithm [1].

Algorithm 12: RNE $(node\_id, QS, result)$

---

input : id of a node; segments within network range e from q in entry of $node\_id$; result
set
if $node\_id$ *is an intermediate node* **then**
  compute QS_i for each entry E_i in $node\_id$;
  for *each entry E_i in $node\_id$* **do**
    if $QS_i \neq \emptyset$ **then**
      $\mid$ RND($E_i.node\_id, QS_i, result$);
    end
  end
else
  $result_{node\_id}$ = plane-sweep($node\_id.entries, QS_i$);
  sort $result_{node\_id}$ to remove duplicates;
  $result = result \bigcup result_{node\_id}$;
end

An alternative is to use the methodology suggested in [12]. In particular, the MBR of all segments in QS is applied as a range query to the object R-Tree. When a leaf node is reached, its contents are joined with QS, using plane sweep. This technique performs a simple intersection test at each visited R-Tree node; however, if the network range is large and irregular, it may visit numerous tree nodes that do not overlap any qualifying segment (e.g., E_1 in Fig. 5.4). Finally, if QS does not fit in memory, the join is performed in a block-nested loops fashion, i.e., RNE is repeatedly applied for subsets of QS that fit in memory and the partial results are materialized. Another approach is to compute all qualifying segments, materialize them, and join them with the object R-Tree using one of the spatial join algorithms that are applicable in the presence of a single tree [17].

5.6 Summary

In this chapter, we introduce the query types for moving objects, such as the NN, range, and density query. We also discuss how to process range and NN queries in a spatial network, based on the Euclidean restriction and network expansion frameworks, covering the most common processing tasks. This provides an introduction to several interesting and practical directions for moving objects querying.

References

1. Arge L, Procopiuc O, Ramaswamy S, Suel T, Vitter JS (1998) Scalable sweeping-based spatial join. In: Proceedings of the 24th international conference on very large data bases (VLDB 1998), New York City, pp 570–581
2. Faloutsos C, Ranganathan M, Manolopoulos Y (1994) Fast subsequence matching in time-series databases. In: Proceedings of the 1994 ACM SIGMOD international conference on management of data (SIGMOD 1994), Minneapolis, pp 419–429
3. Güting R, Bohlen M, Erwig M, Jensen C, Lorentzos N, Schneider M, Vazirgiannis M (2000) A foundation for representing and querying moving objects. ACM Trans Database Syst 25(1):1–24
4. Hadjieleftheriou M, Kollios G, Gunopulos D, Tsotras VJ (2003) On-line discovery of dense areas in spatio-temporal databases. In: Proceedings of the 8th international symposium on advances in spatial and temporal databases (SSTD 2003), Santorini Island, pp 306–324
5. Hjaltason G, Samet H (1999) Distance browsing in spatial databases. ACM Trans Database Syst 24(2):265–318
6. Jensen CS, Lin D, Ooi BC, Zhang R (2006) Effective density queries on continuously moving objects. In: Proceedings of the 22nd international conference on data engineering (ICDE 2006), Atlanta, p 71
7. Kolahdouzan M, Shahabi C (2004) Voronoi-based K nearest neighbor search for spatial network databases. In: Proceedings of the 30th international conference on very large data bases (VLDB 2004), Toronto, pp 840–851
8. Kwon D, Lee SL, Lee S (2002) Indexing the current positions of moving objects using the lazy update R-tree. In: Proceedings of the 3rd international conference on mobile data management (MDM 2003), Singapore, pp 113–120
9. Mokhtar H, Su J (2004) Universal trajectory queries for moving object databases. In: Proceedings of the 2004 IEEE international conference on mobile data management (MDM 2004), Berkeley, pp 133–144
10. Ni J, Ravishankar CV (2007) Pointwise-dense region queries in spatio-temporal databases. In: Proceedings of the 23rd international conference on data engineering (ICDE 2007), Istanbul, pp 1066–1075
11. Papadias D, Zhang J, Mamoulis N, Tao Y (2003) Query processing in spatial network databases. In: Proceedings of the 29th international conference on very large data bases (VLDB 2003), Berlin, pp 790–801
12. Papadopoulos A, Rigaux P, Scholl MA (1999) Performance evaluation of spatial join processing strategies. In: Proceedings of the 6th international symposium on advances in spatial databases (SSD 1999), Hong Kong, pp 286–307
13. Patel JM, Chen Y, Chakka VP (2004) STRIPES: an efficient index for predicted trajectories. In: Proceedings of the 2004 ACM SIGMOD international conference on management of data (SIGMOD 2004), Paris, pp 637–646
14. Pfoser D, Jensen CS (1999) Capturing the uncertainty of moving-object representations, In: Proceedings of the 6th international symposium of SSD 1999, Hong Kong, pp 111–131
15. Pfoser D, Jensen CS (2003) Indexing of network constrained moving objects. In: Proceedings of the 11th ACM international symposium on advances in geographic information systems (GIS 2003), New Orleans, pp 25–32
16. Pfoser D, Jensen CS, Theodoridis Y (2000) Novel approaches in query processing for moving object trajectories. In: Proceedings of the 26th international conference on very large data bases (VLDB 2000), Cairo, pp 395–406
17. Rigaux P, Scholl M, Voisard A (2002) Spatial databases: with application to GIS. Morgan Kaufmann, San Francisco
18. Roussopoulos N, Kelley S, Vincent F (1995) Nearest neighbor queries. In: Proceedings of the 1995 ACM SIGMOD international conference on management of data (SIGMOD 1995), San Jose, pp 71–79

19. Seidl T, Kriegel H (1998) Optimal multi-step K-nearest neighbor search. In: Proceedings of the 1998 ACM SIGMOD international conference on management of data (SIGMOD 1998), Seattle, pp 154–165
20. Song Z, Roussopoulos N (2003) SEB-tree an approach to index continuously moving objects. In: Proceedings of the 4th MDM conference, Melbourne, pp 340–344
21. Tao Y, Papadias D, Sun J, The TPR*-tree: an optimized spatio-temporal access method for predictive queries. In: Proceedings of the 29th VLDB conference, Berlin, pp 790–801
22. Yiu ML, Mamoulis N, Papadias D (2005) Aggregate nearest neighbor queries in road networks. IEEE Trans Knowl Data Eng 17(6):820–833
23. Yiu ML, Papadias D, Mamoulis N, Tao Y (2006) Reverse nearest neighbors in large graphs. IEEE Trans Knowl Data Eng 18(4):540–553

Chapter 6
Moving Objects Advanced Querying

Abstract So far, we have introduced the basic querying for moving objects. There are still some advanced querying for moving objects. It is more difficult to deal with these queries. In this chapter, we introduce a few advanced queries, especially similar trajectory queries and density queries for moving objects. The goal of similar trajectory queries is to find the moving patterns in the trajectories of moving objects, while density queries are to efficiently find dense areas with high concentration of moving objects. We will discuss how to process both the snapshot and continuous density queries in this chapter.

Keywords Spatio-temporal query • Density query • Similar trajectory query • Convoy query • Spatial network • Moving object databases

6.1 Introduction

Recently, many location sensors such as GPS have been developed, and we can obtain the trajectory of users and moving objects using these sensors. Trajectory data are widely used in location-aware systems, transportation navigation systems, and other location-based information systems. These applications have stored within them several trajectories, and these trajectories may include useful individual patterns of each user. For example, by analyzing trajectories of users who work in a building, we can find passages, rooms, stairs, and other facilities that are used frequently. The result of the analysis can be used for the management and maintenance of the buildings. In the case of a navigation system, a driver can check the route to a city by referring to the trajectories of other users who have driven to the city earlier. In another case, we can study movement characteristics to improve performance in a sport by analyzing the motion data measured by the sensors attached to the bodies of top sport players. Thus, similar trajectory queries are produced to find the moving patterns embedded in the trajectories.

X. Meng et al., *Moving Objects Management: Models, Techniques and Applications*, DOI 10.1007/978-3-642-38276-5_6,

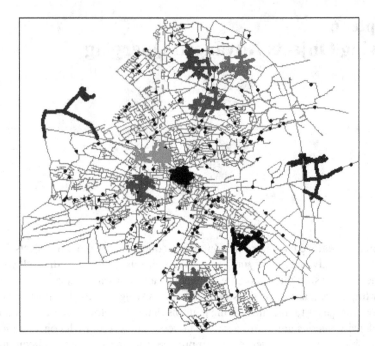

Fig. 6.1 Density query

The distance-based queries such as range queries or NN queries that are defined using the distance between the trajectory of a moving object and an indicated point in a space are useful in location management of moving objects. However, these queries do not have enough power to analyze the pattern of the objects' motion. As mentioned above, because we are interested in the extraction of the individual moving patterns of each object from the trajectories, it is necessary to develop more powerful tools to analyze the trajectories. The similar sequence matching has been studied for many years [1, 2, 4, 13, 15, 25], but the traditional techniques for data sequence including distance function and index cannot be used in the case of the trajectories of moving objects. In this chapter, we present a data model for trajectories of mobile data and a similar trajectory query based on the distance between two trajectories by extending the similarity used in the time series database systems.

Density queries are another type of important queries for moving objects. The objective is to efficiently find dense areas with high concentration of moving objects. Density queries can be used in traffic management systems to identify and predict the congested areas or traffic jams. For example, the transportation bureau may monitor the dense regions periodically in order to identify traffic jams. An instance of density query is shown in Fig. 6.1. The lines depict the road network, points indicate moving objects, and the dense regions are marked in different colors.

Existing studies on density queries [14, 16] assume the objects to be moving in a freestyle and define the density query in the Euclidean space. In this setting, it is difficult to efficiently answer the general density-based queries. The focus is hence turned to simplified queries [14] or specialized density queries without answer loss [16]. These methods use the grid to partition the data space into disjoint cells and report the dense regions with the fixed size. However, the real dense areas may be larger or smaller than the fixed-size rectangle and appear in different shapes. Simplifying the dense query to return the area with fixed size and shape cannot reflect the natural congested area in real-life applications. We focus on density queries in the road- network setting, where the dense area consists of road segments containing large number of moving objects and may be formed in any size and shape. The real congested areas can therefore be obtained by finding the dense segments. In this chapter, we introduce a cluster-based method for monitoring the snapshot of dense areas of moving objects in a road network. Then, we discuss how to *continuously* monitor dense regions for moving objects. Based on the notion of safe interval, we propose effective algorithms to evaluate and keep track of dense regions.

6.2 Similar Trajectory Queries for Moving Objects

Moving object trajectories can be considered as two (X-Y plane)- or three (X-Y-Z plane)-dimensional time series data. In terms of similarity-based queries, we are concerned with the movement shape of the trajectories; sequences of sampled vectors are important in measuring the similarity between two trajectories, and time component is less important so can be ignored. This separates similarity-based retrieval from queries in spatio-temporal databases where time components of trajectories are important to answer time slice or time interval queries [23]. Considerable research has been conducted on similarity-based retrieval on one-dimensional time series data, such as stock or commodity prices, sales volume, weather data, and biomedical measurements. However, the distance functions and indexing methods proposed for one-dimensional time series data cannot be directly applied to moving object trajectories due to their unique characteristics.

- Trajectories are usually two- or three-dimensional data sequences, and a trajectory dataset often contains trajectories with different lengths. Most of the earlier proposals on similarity-based time series data retrieval focused on one-dimensional time series data [1, 6, 18, 20, 28].
- Trajectories usually have many outliers. Unlike stock, weather, or commodity price data, trajectories of moving objects are captured by recording the positions of the objects from time to time (or tracing moving objects from frame to frame in videos). Thus, due to sensor failures, disturbance signals, or errors in detection techniques, many outliers may appear. Longest common subsequences (LCSS) has been applied to address this problem [27]; however, it does not consider

various gaps between similar subsequences, which leads to inaccuracy. The gap refers to a sub-trajectory between two identified similar components of two trajectories.

- Similar movement patterns may appear in different regions of trajectories. Different sampling rates of tracking and recording devices combined with different speeds of the moving objects may introduce local shifts into trajectories (i.e., the trajectories follow similar paths, but certain sub-paths are shifted in time). Even though similarity measures, such as dynamic time warping (DTW) [8,17,29] and edit distance with real penalty (ERP) [7], can be used to measure the similarity between trajectories with local shifts, they are sensitive to noise.

In order to manage trajectories in database systems, we define a data model of trajectories as directed lines in a space, and the similarity between trajectories is defined as the Euclidean distance between directed discrete lines. Our proposed similarity queries can be used to find useful patterns embedded into the trajectories, for example, the trajectories of mobile cars in a city may include patterns for possible traffic jams.

6.2.1 Problem Definition

It is difficult to define the similarity between lines in a space. However, we find some useful clues through study of time series databases [5, 18, 24]. The time series database systems can store time series data such as temperature, economic indicators, population, and wave signals, in addition to supporting queries for extracting patterns from the time series data. Most of the time series database systems adopt the Euclidean distance between two time data sequences [18] for analysis. Since trajectory is a type of time series data, the time series databases can deal with trajectories efficiently. However, trajectory not only has a time series data feature but also has a space feature. For example, it is difficult for the time series database to find data for geographic and spatial queries.

In order to define the similarity between trajectories, it is necessary first to define the trajectory. Hence, we define the data model for the trajectory of moving objects.

A real-world trajectory is a directed continuous line with a start and an end point (Fig. 6.2a). Given a two-dimensional space $\mathbf{R}^2$ and a closed time interval $I_\lambda = [t, t']$ with $t < t'$, a trajectory λ is defined as follows.

Definition 6.1. A *trajectory* is the image of a continuous mapping: $\lambda : I_\lambda \rightarrow \mathbf{R}^2$.

This definition is a temporal extension of the definition of a simple line described in [3]. Next, we denote the length of trajectories in $\mathbf{R}^2$ as L_S and the interval of trajectories in temporal space as L_T.

Definition 6.2. The *length of trajectory* λ during a period $[t_0, t_1]$ is denoted as $L_S(\lambda, [t_0, t_1])$ calculated as follows:

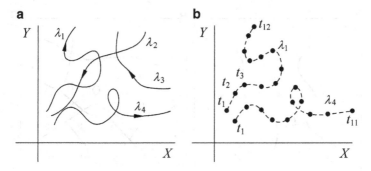

Fig. 6.2 Trajectory of moving objects. (**a**) Trajectory in the real world. (**b**) Trajectory stored in a database

$$L_S(\lambda, [t_0, t_1]) = \int_{t_0}^{t_1} \sqrt{(dx/dt)^2 + (dy/dt)^2} dt, \text{where } \lambda(t) = (x, y) \qquad (6.1)$$

The length of the whole trajectory is denoted as $L_S(\lambda) = L_S(\lambda, [t, t'])$.

Definition 6.3. Given that the $\mathbf{x} = (x, y)$ is a vector in space $\mathbf{R}^2$, the *temporal interval of trajectory* λ between $\mathbf{x}_i$ and $\mathbf{x}_j$ on λ is defined as follows:

$$L_T(\lambda, [\mathbf{x}_i, \mathbf{x}_j]) = |t_j - t_i|, \text{where } \lambda(t_i) = \mathbf{x}_i, \lambda(t_j) = \mathbf{x}_j, \text{and } t_i, t_j \in I_\lambda[t, t'] \qquad (6.2)$$

$$L_T(\lambda) = |t' - t| \qquad (6.3)$$

However, a positioning device such as GPS does not continuously measure the coordinates of a moving object, but samples such data. The measured data are thus a sequence of coordinates of positions shown in Fig. 6.2b. Hence, we define discrete trajectory $\dot{\lambda}$ as a discrete function. Each vector $\mathbf{x}_i$ represents a position of a moving object at each time $\mathbf{T}_{\dot{\lambda}} = \{t_0, t_1, \ldots, t_m\}$ in the space.

Definition 6.4. A *discrete trajectory* is the image of a discrete mapping: $\dot{\lambda}$: $\mathbf{T}_{\dot{\lambda}} \to \mathbf{R}^2$.

A discrete trajectory can be represented as a vector sequence $< \mathbf{x}_{t1}, \ldots, \mathbf{x}_{tm} >$ as well. If $\mathbf{T}_{\dot{\lambda}} = \{1, 2, \ldots, m\}$, we denote the discrete trajectory $\dot{\lambda}$ as just a simple vector sequence $< \mathbf{x}_1, \ldots, \mathbf{x}_m >$. Additionally, where $\dot{\lambda}(t_i) = \mathbf{x}_i$, we introduce several notations: $\mathbf{T}_{\dot{\lambda}}(i) = t_i$, $\mathbf{X}_{\dot{\lambda}}(i) = \mathbf{x}_i$, and $|\dot{\lambda}|$ is the number of the vectors included in $\dot{\lambda}(|\dot{\lambda}| = |\mathbf{T}_{\dot{\lambda}}|)$. Next, we define the distance between two vectors $\mathbf{x}, \mathbf{x}'$ in $\mathbf{R}^2$.

Definition 6.5. The *distance of vectors* $\mathbf{x}, \mathbf{x}'$ is defined as:

$$D(\mathbf{x}, \mathbf{x}') = \sqrt{(x - x')^2 + (y - y')^2} \qquad (6.4)$$

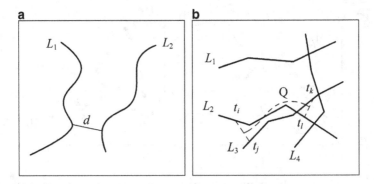

Fig. 6.3 Distance between trajectories. (**a**) EU distance based trajectory similarity. (**b**) Shape based trajectory similarity

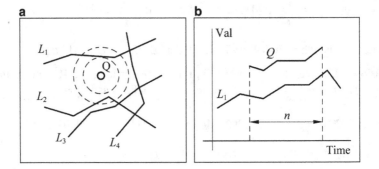

Fig. 6.4 Existing kNN approaches. (**a**) Previous spatial kNN. (**b**) Previous temporal kNN

6.2.2 Trajectory Similarity

In time series databases, the similarity between two sets of time series data is typically measured by the Euclidean distance [5, 18], which can be calculated efficiently. However, there have been few discussions on the similarity between two lines in space because the previous approaches for spatial queries have focused on the "distance" between a point and a line [9, 10, 19]. The aim of the previous approaches is mainly to find objects that pass a point near the indicated point, such as a car passing through a street. On the other hand, we are concerned with the "shape" of the trajectory. In order to calculate shape-based similarities among trajectories, it is necessary to define a new similarity for the trajectories, as shown in Fig. 6.3b.

In general, the similarity query is represented as a kNN query [9, 19]. There are two types of existing approaches: one is based on spatial similarities, and the other is based on similarity between two time series data. The example of the existing spatial kNN query is illustrated in Fig. 6.4a. In this case, the answer is L_1, L_2 when k is 2. On the other hand, the similarity between two time series data is defined as the Euclidean distance between two time series, where the length of each is n. The distance is defined as the Euclidean distance between two n-dimensional vector data [18] shown in Fig. 6.4b. While this distance of the time series data is based on shape, the distance is defined only in the case of $\mathbf{R}^1 \times \mathbf{T}(\mathbf{T} = [0, \infty])$, but not in the case of $\mathbf{R}^n \times \mathbf{T}$, shown in Fig. 6.3b. Since the trajectory has both spatial and temporal features, we consider three types of similarity queries for trajectories as follows:

- **Spatio-temporal similarity**: based on a spatio-temporal feature in $\mathbf{R}^2 \times \mathbf{T}$
- **Spatial similarity**: based on a spatial only feature in $\mathbf{R}^2$ without temporal features
- **Temporal similarity**: Based on a temporal only feature in $\mathbf{R}^1 \times \mathbf{T}$ without spatial features

As mentioned above, the trajectory has a time series data feature. We define the similarity between two trajectories in the same manner as for the similarity defined in the time series query [18]. For the time series database, the similarity of the two time series data, where each has n values, is given by the Euclidean distance between vectors in $\mathbf{R}^n$. In [5] and [18], when there are two time series data, $c =< w_1, w_2, \ldots, w_n >, c' =< w'_1, w'_2, \ldots, w'_n >$, the distance $D(c, c')$ is defined as follows:

$$D(c, c') = \sqrt{(w_1 - w'_1)^2 + \cdots + (w_n - w'_n)^2} \qquad (6.5)$$

This definition can be extended if each vector $\mathbf{x}$ is a vector in space $\mathbf{R}^2$, when the time series vectors are $\mathbf{X} =< \mathbf{x}_1, \mathbf{x}_2, \ldots, \mathbf{x}_n >, \mathbf{X}' =< \mathbf{x}'_1, \mathbf{x}'_2, \ldots, \mathbf{x}'_n >$, and the distance is $D(c, c')$. We define the distance between two time series vectors $D(\mathbf{X}, \mathbf{X}')$ by extending the definition of $D(c, c')$, as follows:

$$D(\mathbf{X}, \mathbf{X}') = \sqrt{D(\mathbf{x}_1 - \mathbf{x}'_1)^2 + \cdots + D(\mathbf{x}_n - \mathbf{x}'_n)^2} \qquad (6.6)$$

The defined distance $D(\mathbf{X}, \mathbf{X}')$ can be used only in the case where each vector $\mathbf{x} \in \mathbf{X}$ is measured by the same interval, that is, $\Delta t = t_{i+1} - t_i (i = 1, \ldots, n - 1)$, where t_i is an interval from the time when $\mathbf{x}_i$ is measured. However, each vector in the trajectory is not always measured by the same interval Δt because positioning devices often lose the data. Therefore, to calculate the similarity using our definition, we define a temporal normalized discrete trajectory $\dot{\lambda}_{\Delta t}$ for trajectory λ, as follows:

Definition 6.6. Given a trajectory λ defined for time interval $[t_S, t_E]$ and a natural number m, the *temporal normalized discrete trajectory* $\dot{\lambda}_{\Delta t}$ is defined as follows:

$$\dot{\lambda}_{\Delta t} =< \lambda(t_S), \lambda(t_S + \Delta t), \ldots, \lambda(t_S + m\Delta t) >, \text{where } t_S + m\Delta t = t_E \quad (6.7)$$

Intuitively, this discrete trajectory $\dot{\lambda}_{\Delta t}$ is the re-sampled trajectory per fixed interval Δt from λ. In other words, $\dot{\lambda}_{\Delta t}$ is generated by dividing λ into equal interval Δt. For discrete trajectory $\dot{\lambda}$, we can use the piecewise linear approximation $\tilde{\lambda}$ instead of λ.

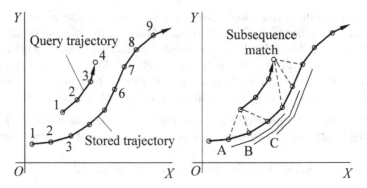

Fig. 6.5 Similarity query for trajectories

Definition 6.7. Given two trajectories λ and λ' with the same temporal length (i.e., $L_T(\lambda) = L_T(\lambda')$) and a natural number m, the *spatio-temporal distance* (*similarity*) $D_{TS}(\lambda, \lambda')$ between λ and λ' is defined as follows:

$$D_{TS}(\lambda, \lambda') = \frac{1}{m+1} \sqrt{\sum_{i=0}^{m} D(\mathbf{X}_{\lambda_{\Delta t}}(i), \mathbf{X}_{\lambda'_{\Delta t}}(i))^2}, \text{ where } \Delta t = \frac{L_T(\lambda)}{m} = \frac{L_T(\lambda')}{m}$$

$$(6.8)$$

Note that $D_{TS}(\dot{\lambda}, \dot{\lambda'})$ can be defined as $D_{TS}(\tilde{\lambda}, \tilde{\lambda'})$. In this definition, the similarity is the Euclidean distance between trajectories represented as $m + 1$-dimensional vectors, and the interval of each trajectory is normalized. Using this definition, it is possible to find trajectories whose shape is more similar to the query trajectory than that which can be found using previous methods.

6.2.3 Query Processing

Based on these definitions, we consider the shape-based similarity query for trajectories. Here, $\dot{\Lambda}$ is the set of discrete trajectories stored in the database, and each $\dot{\lambda}_i(\dot{\lambda}_i \in \dot{\Lambda})$ is a discrete trajectory, such as $\dot{\lambda}_i = <\mathbf{x}_1, \mathbf{x}_2, \dots, \mathbf{x}_m>$. The query trajectory $\dot{\lambda}_q$ is given as $\dot{\lambda}_q = <\mathbf{x}_1, \mathbf{x}_2, \dots, \mathbf{x}_n>$. The shape-based range query can then be defined using $\dot{\Lambda}, \lambda_q$, and the previous defined distance between two time series vectors, as follows:

Definition 6.8. The process for calculation of the shape-based range query $Q_{range}(\theta, \dot{\lambda}_q, \dot{\Lambda})$ is given in Algorithm 13. The range query is defined as a subsequence match of trajectories as shown in Fig. 6.5.

Algorithm 13: $Q_{range}(\theta : integer, \dot{\Lambda}, \dot{\lambda}_q) : \dot{\Lambda}_a$

Input: $\dot{\Lambda}, \dot{\lambda}_q, \theta (\theta$ is a natural number)
Output: $\dot{\Lambda}_a, \{\dot{\lambda}_{a1}, \ldots, \dot{\lambda}_{ak}\} \in \dot{\Lambda}_a$
begin
    $l = |\dot{\lambda}_q|; \dot{\Lambda}_a = \phi;$
    foreach $\dot{\lambda}_i$ *in* $\dot{\Lambda}$ **do**
        for $j = 1$ *to* $|\dot{\lambda}_i| - l + 1$ **do**
            $\dot{\lambda}_{ij} = subsequence(\dot{\lambda}_i, j, l);$
            //This function will return a subsequence of the original sequence $\dot{\lambda}_i$, such as
            $< \mathbf{x}_j, \mathbf{x}_{j+1}, \cdots, \mathbf{x}_{j+l-1} >$, each $\mathbf{x} \in \dot{\lambda}_i;$
            if $D(\dot{\lambda}_q, \dot{\lambda}_{ij}) < \theta$ **then**
                | Add $\dot{\lambda}_{ij}$ to $\dot{\Lambda}_a;$
            end
        end
    end
    return $\dot{\Lambda}_a$
end

In addition, the nearest neighbor query can be defined using the distance between trajectories. In our definition, the temporal features are not indicated in the query; however, we consider that the temporal features can be indicated independently from the range query. For example, a query "$Q_{range}(\theta, \dot{\lambda}_q, \dot{\Lambda}) \wedge 11 : 00 < T_{\dot{\lambda}_{ai}}(1) < 12 : 00$" involves retrieving subsequences $\dot{\lambda}_{ai}$ where the distance between $\dot{\lambda}_q$ and $\dot{\lambda}_{ai}$ is less than θ. Moreover, the first vector in $\dot{\lambda}_{ai}$ is measured within the interval [11:00, 12:00].

6.3 Convoy Queries on Moving Objects

Finding a group of objects that are likely to travel together is meaningful for many applications. Given a set of objects' trajectories 0, density constraints m and e, and a lifetime k, a convoy query retrieves all convoys, each of which has at least m objects that traveled closely with respect to a distance c during at least k consecutive time stamps. Discovering convoys in a large database involves a large number of combinations for processing *spatio-temporal join*. However, join queries lead to more expensive computations. This section presents three effective algorithms for answering the convoy query. One method is based on a solution for finding moving clusters and modified for the problem of convoy discovery. The other two methods reduce the number of vertices of original trajectories by using line simplification algorithms and then find convoys over the simplified trajectories.

6.3.1 Spatial Relations Among Convoy Objects

To address the issue of identifying clusters, density-based notions [12] of clustering are used to explain the spatial closeness among the convoy objects. The neighborhood of a point p is denoted by $N_e(p)$ and is defined by $N_e(p) = \{q \in S \mid D(p,q) \le e\}$, where S is a given dataset and $D(p,q)$ is a Euclidean distance between p and q.

A point p is *directly density reachable* from a point q w.r.t. distance e and minimum number of points m, if $P \in N_e(q)$ and $|N_e q| \ge m$.

A point p is *density reachable* from a point q w.r.t. e and m, if there is a chain of points $p_1, p_2, \cdots, p_n, p_1 = q, p_n = p$ such that p_{i+1} is directly density reachable from p_i.

A point p is *density connected* to a point q w.r.t. e and m, if there is a point x such that both p and q are density reachable from x w.r.t. e and m.

6.3.2 Coherent Moving Cluster (CMC)

A convoy is necessarily discovered from a sequence of consecutive snapshot clusters if the number of identical objects among the clusters exceeds a given m. The algorithm is shown as follows:

1. The verification of moving cluster is done by checking if two snapshot clusters at two consecutive time stamps have a larger or equal percentage of common objects to a given threshold θ ($|\frac{c_t \cap c_{t+1}}{c_t \cup c_{t+1}}| \ge \theta$), where c_t and c_{t+i} denote two adjacent snapshot clusters at time t and $t + 1$, respectively. Instead of using θ, CMC modifies the verification process to $|c_t \cap c_{t+1}| \ge m$, which is a stricter constraint of checking identical objects.
2. A moving cluster can be formed as long as only two snapshot clusters satisfy the θ constraint. Hence, a cluster can be checked if it lasts for at least k time stamps.

6.3.3 Convoy Over Simplified Trajectory (CoST)

In CoST, trajectories are simplified by using the Douglas-Peucker algorithm (DP) [11], and then apply a clustering method on simplified trajectory segments instead of actual points. To avoid false dismissals of incorrect query results produced by trajectory simplification, a safe bound of errors is determined by providing a lower bounding lemma. Let O'_1, O'_2 be two simplified trajectories of corresponding original trajectories O_1, O_2, respectively. Let $D_{pp}(p,q)$ be the Euclidean distance from a point $p \in O_1$ to another $q \in O_2$, $D_{pp}(p',q')$ be the distance between the corresponding points $p' \in O'_1$ to $q' \in O'_2$, and δ be a tolerance value for DP. Then:

Fig. 6.6 Adaptive sweep line processing

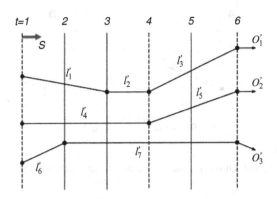

Lemma 6.1. $D_{pp}(p', q') \le 2\delta + D_{pp}(p, q)$

The lemma means that distance between two points of simplified trajectories is bounded by $2\delta+$ the Euclidean distance between the corresponding points on original trajectories. Now, another lemma is established for density clustering over simplified trajectories. Let $l'_1 = \{p'_1, p'_j\}$ be a trajectory segment on a simplified trajectory O'_1 and $L_1 = \{p_i, p_{i+1}, \cdots, p_j\}$ be a polyline during the time interval $[i, j]$ on the corresponding original trajectory O_1. Likewise, let $l'_2 = \{q'_u, q'_v\}$ be a trajectory segment on O'_2, the corresponding $L2 = \{q_u, q_{u+1}, \cdots, q_v\}$ of O_2. If the distance between trajectory segments with the shortest distance is measured as Fig. 6.7a, the following lemma applies.

Lemma 6.2. *If $p \in L_1$ is density connected to $q \in L_2$ w.r.t. e and m, l'_1 is density connected to l'_2 w.r.t. $2\delta + e$ and m.*

Since $D_{LL}(l'_1, l'_2)$ represents any distance between a point pair form l'_1 and l'_2 as the shortest one, the points are more likely to be within the range e of clustering. As a result, though a cluster c' on simplified trajectories may contain some objects that do not belong to the corresponding cluster c on original trajectories, it never misses any objects of c on c'. This guarantees no false dismissal for query answers.

The procedure of query processing is as follows: the clustering computation is skipped until the total number of objects is found. Consider Fig. 6.6 that describes three simplified trajectories. Let S_t be a set of trajectory segments found until the sweep line s arrives at time t. When $s = 2$, l'_6 is found and l'_1 is detected at $s = 3$. So far, only two objects O'_1 and O'_3 are found, so this sweeping process goes on until O'_2 is found at $s = 4$. Now the clustering method is applied on $S_4 = \{l'_1, l'_2, l'_3, l'_4, l'_6, l'_7\}$. The same process is repeated until the end of time stamps. This query processing scheme achieves high efficiency of the convoy discovery by performing the convoy validation between two consecutive clusters.

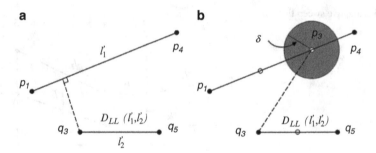

Fig. 6.7 Different distance measures of trajectory segments

6.3.4   Spatio-Temporal Extension (CoST*)

Though the CoST method does not have a false negative by bounding errors, it
may report some false positives which need further refinement steps of the query
processing. Due to the expensive computational cost of the refinement step, the error
bound is tighten to reduce the number of false positives by applying a different
line simplification method and another distance measure of trajectory segments.
Unlike the original DP algorithm using perpendicular distance, its spatio-temporal
extension in [21], say DP*, considers the ratio of time. This fact brings a less
reduction ratio of vertices on original trajectories, since the perpendicular distance
for approximation is the shortest distance. Thus, query processing needs to perform
clustering more often, which involves lower efficiency.

On the other hand, DP* may bring stronger filtering power for query processing
by larger distance measures of trajectory segments. Because DP* computes the ratio
of time when it approximates trajectories, an omitted point after the simplification is
trackable by the combination of a liner interpolation and δ. In Fig. 6.7b, for example,
p_3 was inside the gray circle having d radius before applying DP* method. Hence,
the distance between the trajectory segments can be measured by $D_{LL} * (L'_1, L'_2) =
D_{pp}(p_3, q_3)$, which is larger than $D_{LL}(L'_1, L'_2)$ in Fig. 6.7a. Such larger distances
have higher probabilities to be out of the range for clustering, thus this approach
produces less numbers of false positives.

For query processing, an enhanced method, called CoST*, is developed to
replace two components of the CoST method. (1) It applies DP* for the trajectory
simplification. (2) DLL* is used for the distance measure of the density clustering,
instead of D_{LL}. CoST* produces less numbers of false positives by the close-fitting
error bound, thus it brings higher efficiency of the discovery process.

6.4   Density Queries for Moving Objects in Spatial Networks

The issue of density queries for moving objects was first proposed in [14]. The objective is to find regions in space and time with the density higher than a given threshold. In paper [14], the authors find the general density-based queries difficult to be answered efficiently and hence turn to simplified queries. Specifically, they partition the data space into disjoint cells and simplified density query report cells, instead of arbitrary regions that satisfy the query conditions. This scheme may result in answer loss. To solve this problem, Jensen et al. [16] define an effective density query to guarantee that there is no answer loss. Both studies assume the objects to be moving in a freestyle and define the density query in Euclidean space. However, efficient dynamic density query in spatial networks is more crucial for many applications. Consider this real-world example: in the case of queries related to vehicle distribution in the road network, users would like to know real-time traffic density distribution. Clearly, in this case the Euclidean density query methods are inapplicable, since the path between two cars is restricted by the underlying road network. Additionally, these existing query methods cannot reflect the natural dense area in a road network since they simplify the density query to return the area with fixed size and shape. Grid-based algorithms also ignore the network constraint and result in inaccurate query results. It is natural to represent the dense area in a road network as road segments containing large number of moving objects. Considering the feature of road networks, we will introduce a cluster-based density querying algorithm.

6.4.1   Problem Definition

As the result of density queries in the road network is a set of dense segments, we first introduce the concepts of *density* and *dense segment*.

Definition 6.9. The *density* of a road segment s is represented as $density(s) = N/len(s)$, where N is the number of objects on s and $len(s)$ is the length of s.

Definition 6.10. The road segment s is a *dense segment (DS)* if and only if $density(s) \geq \rho$, where ρ is a user-defined density parameter.

A straightforward method to process the query is to traverse all objects moving on a road network to compute dense regions by their number, the length of the segment, and a given density threshold. Figure 6.8 shows a density query in a road network. Obviously, the cost is very high and it is difficult to obtain effective results. Specifically, the following three issues are likely to be encountered in the case of the query results:

1. Different DS may be overlapped, such as Case 1 in Fig. 6.8.

Fig. 6.8 An example of
density query

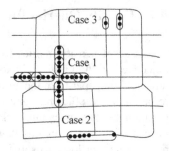

2. The distribution of moving objects may be very skewed in some DS, i.e., the
   distribution of objects is dense in one part of a DS, but it is sparse in another part,
   such as Case 2 in Fig. 6.8.
3. Some DS may contain very few objects, such as Case 3 in Fig. 6.8.

Such query results are less useful. Thus, we define an *effective density query* in
a road network to find the useful dense regions with a high concentration of objects
and symmetrical distribution of objects as well as no overlaps.

Definition 6.11. Given density parameter ρ, *effective road-network density query*
(*e-RNDQ*) aims to find all dense segments that satisfy the following conditions:

1. Any dense segment set cannot be intersecting (namely, no overlaps).
2. In each dense segment set, the distance between any neighboring object is not
   more than a given distance threshold δ.
3. The length of dense segments is not less than a given length threshold L.
4. Any dense segment containing moving objects is in the query result set.

The first condition ensures that the result is not redundant. It avoids the Case 1 in
Fig. 6.8. The second condition guarantees that objects are symmetrically distributed
in a dense segment set. The third condition provides the restriction that there are
no small segments that only contain few objects in the result. The fourth condition
ensures that query results do not suffer from answer loss.

6.4.2   Cluster-Based Query Preprocessing

To reduce the cost of clustering maintenance, we introduce the definition of *cluster
unit* (CU). A cluster unit is a group of moving objects close to each other at present
and near-future time. The cluster unit will be incrementally maintained according to
the moving objects within it. Specifically, we constrain the objects in a cluster unit
moving in the same direction and on the same segment. For keeping the objects in
a cluster unit dense enough, the network distance between each pair of neighboring
objects in a cluster unit should not exceed a system threshold ϵ. As mentioned

earlier, we assume that objects move in a piecewise linear manner and the next segment to move is known in advance. Formally, a cluster unit is defined as follows:

Definition 6.12. A *cluster unit* is represented by $(O, n_a, n_b, head, tail, ObjNum)$, where O is a list of objects $\{o_1, o_2, \ldots, o_i, \ldots, o_n\}$, $o_i = (oid_i, n_a, n_b, pos_i, speed_i, next\_node_i)$, where pos_i is the relative location to n_a, $speed_i$ is the moving speed, and $(n_b, next\_node)$ is the next segment to move. Without loss of generality, assuming $pos_1 \leq pos_2 \leq \ldots \leq pos_n$, it must satisfy $|pos_{i+1} - pos_i| \leq \epsilon \ (1 \leq i \leq n - 1)$. Since all objects are on the same segment (n_a, n_b), the position of the CU is determined by an interval $(head, tail)$ in terms of the network distance from n_a. Thus, the length of the CU is $|tail - head|$. $ObjNum$ is the number of objects in the CU.

Initially, based on the definition, a set of CUs are created by traversing all segments in the network and their associated objects. The CUs are incrementally maintained after their creation. As time elapses, the distance between adjacent objects in a CU may exceed ϵ. Thus, we need to split the CU. A CU may also merge with its adjacent CUs when they are within the distance of ϵ. Hence, for each CU, we predict the time when they may split or merge. The predicted split and merge events are then inserted into an event queue. Subsequently, when the first event in the queue takes place, we process it and update the affected CUs. This process is continuously repeated. The key challenges are: (1) how to predict split/merge time of a CU and (2) how to process a split/merge event of a CU.

The split of a CU may occur in two cases. The first one is when the CU arrives at the end of the segment (i.e., an intersection node of the road network). When the moving objects in a CU reach an intersection node, the CU has to be split since they may head in different directions. Split time refers to the time when the first object in the CU arrives at the node. In the second case, the split of a CU occurs when the distance between some neighboring objects moving on the segment exceeds ϵ. However, it is not easy to predict the split time since the neighborhood of objects changes over time. Therefore, the main task is to dynamically maintain the order of objects on the segment. We compute the earliest time instance when two adjacent objects in the CU meet at t_m. We then compare the maximum distance between each pair of adjacent objects with ϵ until t_m. If this distance exceeds ϵ at some time, the process stops and the earliest time exceeding ϵ is recorded as the split time of CU. Otherwise, we update the order of objects starting from t_m and repeat the same process until some distance exceeds ϵ or one of the objects arrives at the end of the segment. When the velocity of an object changes over the segment, we need to re-predict the split and merge time of the CU.

To reduce the processing cost of splitting at the end of segment, we propose the group split scheme. When the first object leaves the segment, we split the original CU into several new CUs according to the objects' directions (which can be implied by $next\_node$). On the one hand, we compute a *to-be-expired time* (i.e., the time until the departure from the segment) for each object in the original CU and retain the CU until the last object leaves the segment. On the other hand, we attach a *to-be-valid time* (with the same value as the *to-be-expired time*) for each object in the new CUs. Only valid objects will be considered while constructing CUs.

The merge of CUs may occur when adjacent CUs in a segment are moving together (i.e., their network distance $\leq \epsilon$). To predict the initial merge time of CUs, we dynamically maintain the boundary objects of each CU and their validity time (the period when they are treated as boundary of the CU) and compare the minimum distance between the boundary objects of two CUs with the threshold ϵ at their validity time. The boundary objects of CUs can be obtained by maintaining the order of objects during computation of the split time.

The processing of the merge event is similar to the split event on the segment. We get the merge event and time from the event queue to merge the CUs into one CU and compute the split time and merge time of the merged CU. Finally, the corresponding affected CUs in the event queue are updated.

Besides the split and merge of CUs, new objects may come into the network or existing objects may leave. For a new object, we locate all CUs of the same segment that the object enters and check whether the new object can join any CU according to the CU definition. If the object can join some CU, its split and merge events are updated. If no such CUs are found, a new CU for the object is created and the merge event is computed. For a leaving object, we update the split and merge events of its original CU if necessary.

6.4.3   Density Query Processing

Based on the dynamic CUs, density queries at any time point can be processed efficiently to return dense areas in the road networks. The dense segment we defined in Sect. 6.4.1 is represented as (CU, n_a, n_b, *startpos*, *endpos*, *len*, N), where CU is the set of cluster units on segment (n_a, n_b), *startpos* is the start position of the DS, *endpos* is the end position of the DS, *len* is the length of DS, and N is the number of objects. To obtain the effective dense areas restricted in the e-RNDQ, we introduce the parameter δ to DS.

Definition 6.13. A DS is δ-*DS* if and only if the distance between any adjacent CUs is not more than δ (this guarantees that the distance between any two adjacent objects satisfies $Distance(o_i, o_{i+1}) \leq \delta$) and density is not less than ρ. (For convenience, we abbreviate the term δ-*DS* to DS in the rest of this chapter.)

In fact, δ is a user-defined parameter of the density query and ϵ is a system parameter to maintain the CUs. Since the distance of adjacent objects is not more than ϵ in a CU, in order to retrieve dense areas based on CUs, we require $\epsilon \leq max\{\delta, \frac{1}{\rho}\}$. In the road network, a dense area is represented as a dense segment set, which may contain several DSs in different segments. Therefore, we leverage network nodes to optimize the combination of these DSs.

Definition 6.14. In each DS, n_a is δ-cluster node (δ-CN) of the DS if and only if $|$ startpos-n_a $|\leq \delta$; n_b is δ-CN of the DS if and only if $|$ endpos-n_b $| \leq \delta$.

Definition 6.15. A dense segment set (*DSS*) consists of different DSs where the distance between adjacent DSs is not more than δ, the total length of DSs in the DSS is not less than L, and the density in the DSS is not less than ρ.

Actually, DSS may contain DSs located in different segments where DSs are joined by δ-CN. DSS constitutes the road-network density query results. Suppose the density query parameter is given as (ρ, δ, L, t_q), where t_q is the query time. For query processing based on CUs, our algorithm includes two steps:

1. **The filtering step**: Merge CUs into DSs by checking the parameters ρ and δ, which can prune some unnecessary segments. In this step, we can obtain a series of dense segments, specifically, a list of DSs and δ-CNs.
2. **The refinement step**: Merge the adjacent DSs around δ-CNs to construct the DSS by checking the parameters ρ, δ, L and finally find out the effective density query result consisting of dense segment sets.

Algorithm 14: *Filter(ρ, δ, t_q)*

Input: density threshold ρ, query time t_q
begin
    foreach $e(n_x, n_y)$ *of edgeList* **do**
        if $e.cuList \neq null$ **then**
            create a new DS: ds;
            $cu \leftarrow getFirstCU(e)$;
            $ds.addCU(cu)$; $ds.startpos = cu.pos$;
            if $ds.startpos < \delta$ **then**
                $ds.putCN(n_x)$; δ-$CN[n_x].putDS(ds)$;
            end
            while $getNextCU(e) \neq null$ **do**
                $nextcu \leftarrow getNextCU(e)$;
                if $Dd(ds, nextcu) > \delta$ **or**
                $Dens(ds, nextcu) < \rho$ **then**
                    $ds.endpos = cu.pos + cu.len$; $e.addDS(ds)$;
                    create a new DS: ds;
                    $ds.startpos = nextcu.pos$;
                end
                $ds.addCU(nextcu)$; $cu = nextcu$;
            end
            $ds.endpos = cu.pos + cu.len$;
            if $1 - ds.endpos < \delta$ **then**
                $ds.putCN(n_y)$; δ-$CN[n_y].putDS(ds)$;
            end
            $e.addDS(ds)$;
        end
    end
end

Fig. 6.9 An example to
construct DS and DSS

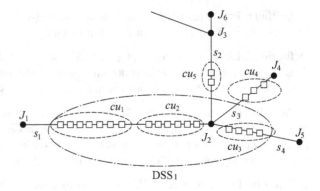

$$DSS_1$$

We explain the two steps of density query processing in detail. First, according
to the network expansion approach [22], we traverse each segment to retrieve CUs
sequentially and then compute the distance between adjacent CUs and their density.
If the distance is not more than δ and the density is not less than ρ, the CUs are
merged to form a DS. Figure 6.9 shows an example. Given $\rho = 1.5$ and $\delta = 2$,
we compute DS at query time t_q. The road segment s_1 (represented as $< J_1, J_2 >$)
includes two CUs named cu_1 and cu_2. Assume that the distance between cu_1 and
cu_2 is 1.2 at t_q, which is less than δ, and the density is 1.8 after merging cu_1 with
cu_2, which is more than ρ, and therefore cu_1 and cu_2 can construct a DS (we call
it DS_1). The start position of DS_1 is the head of cu_1 and the end position of DS_1 is
the tail of cu_2. The number of objects in DS_1 is the sum of the number of objects
in cu_1 and in cu_2. Assume that the distance between DS_1 and node J_2 is 1.0, which
is less than δ, and J_2 is the δ-CN of DS_1 (we call it δ-CN_1). We insert DS_1 into the
DS list of δ-CN_1. In this way, we can obtain DS_2 on s_3 including cu_4 and DS_3 on
s_4 including cu_3. The δ-CN of DS_2 (δ-CN_2) is J_4 and that of DS_3 is J_2. Thus, the
DS list of δ-CN_1 includes DS_1 and DS_3, while the DS list of δ-CN_2 includes DS_2.
Algorithm 14 shows the pseudo-code.

In the refinement step, we compute dense segment sets so that the effective dense
areas can be obtained. We traverse the list of each δ-CN and evaluate whether those
DSs around the δ-CN can construct DSS based on Definition 6.15. For example,
in Fig. 6.9, $L = 100$. As the $Distance(DS_1, \delta\text{-}CN_1)$=1.0 and $Distance(DS_3,$
δ-CN_1)=0.7, the distance between DS_1 and DS_3 is 1.7, which is less than δ. In
addition, if DS_1 is merged with DS_3, the density is more than ρ. Therefore, DS_1
and DS_3 can be merged to form a DSS named DSS_1. In the same way, we check if
there are other dense segments that can be merged with DSS_1 by utilizing its δ-CN
and insert it into DSS_1. Finally, we check if the total length of DSS_1 is more than
L. If so, DSS_1 is one of the answers of the density query. This process is repeated
until all δ-CNs containing dense segments are accessed. Then, we can obtain all
dense areas that are represented as dense segment sets at t_q. Note that a DS may be
involved in the lists of two δ-CNs. To avoid scanning the same nodes repeatedly, we
mark the scanned δ-CN as accessed node. Algorithm 15 shows the pseudo-code of
the refinement step.

Algorithm 15: $Refinement(\rho, \delta, L, t_q)$

Input: density threshold ρ, length threshold of *DSS* L
Output: *Result*: The set of DSSs
begin

    foreach δ-CN_i *of* δ-*CNList* **do**

        if $(\delta$-$CN_i.dsList \neq null)$ *and* $(not\ \delta$-$CN_i.accessed)$ **then**

            /*Q is a priority queue to store all *DSs* around δ-CN_i*/;

            /*δ-Q is a priority queue to store all unaccessed δ-*CNs**/;

            $Q \leftarrow null$; δ-$Q.put(\delta$-$CN_i)$;

            while δ-$Q \neq null$ **do**

                $cn = \delta$-$Q.pop()$; $cn.accessed = true$;

                $Q.addDSs(cn)$; /*add all DSs around cn and sorted*/;

                create a new DSS: dss;

                $ds = Q.pop()$; $dss.addDS(ds)$;

                δ-$Q.putdscn(ds)$; /*add all unaccessed δ-CN around ds*/;

                while $Q \neq null$ **do**

                    $nextDS = Q.pop()$;

                    if $Dist(dss, nextDS) \leq \delta$ and $Dens(dss, nextDS) \geq \rho$ **then**

                        $dss.addDS(nextDS)$;

                        δ-$Q.putdscn(nextDS)$;

                  end

            end

        end

        if $dss.len > L$ **then**

            $Result.insert(dss)$;

    end

    end

    return *Result*;

end

6.5   Continuous Density Queries for Moving Objects

Although many studies have been done on density queries for moving objects, they all focused on how to answer snapshot density queries, where the results are found based on a snapshot of the location dataset. In this section, we focus on continuously monitoring dense regions for moving objects in a highly dynamic environment where the density regions may be changed with location updates of the moving objects. Continuous density query is an important research but has received attention only recently [16]. We provide a definition of continuous density queries for moving objects, which returns useful answers and is amenable to efficient computation. Furthermore, we propose the notion of safe interval for dense/sparse regions to support efficient processing of continuous density queries.

6.5.1 *Problem Definition*

We assume that a collection of objects are moving on the space under consideration, where each object is capable of transmitting its location and velocity to the central server. The central server can predict the object positions based on the location and velocity information and continuously answer density queries. When an object changes its velocity, it updates the new velocity to the central server.

Definition 6.16. A continuous density query returns all the regions that satisfy the following three conditions:

1. The density of the region is not less than ρ.
2. The minimum area of our interest is s and any subarea of the region with an area larger than s must be dense.
3. No two regions in the result set overlap with each other.

Conditions 1 and 2 indicate that each dense region must have more than $\rho \cdot s$ objects. Condition 3 is provided to simplify the search of dense regions, as in the previous study.

We use the TPR-tree to index the moving objects [26]. In the TPR-tree, the position of a moving object is represented by a vector including the reference position and the velocity – $(p(t_{ref}), v)$. We can predict the future location at time t using the following formula:

$$p(t) = p(t_{ref}) + v \cdot (t - t_{ref}) \tag{6.9}$$

In order to find local dense regions, we recursively partition the space by a Quad-tree. The Quad-tree is used to store the state (i.e., dense or sparse) of a subspace, as well as the validity in time, which we call safe interval of the subspace. Thus, a node in the Quad-tree is represented as $((row, col), level, state, safe\_interval)$, where (row, col) is an index to identify the node and *level* denotes the level of the tree that the node belongs to. If the node is a leaf, the *state* can be 0 or 1, which indicates that the region represented by the node is sparse or dense. For a non-leaf node, the *state* can be 0, 1, or 2, where 0 indicates all its children nodes are sparse, 1 indicates all its children nodes are dense, and otherwise, the *state* is 2. The *safe_interval* is the valid time of the state, which is formally defined as follows.

Definition 6.17. The *safe interval* is the time period for which the region remains in its current state. For example, if the region is dense, it will remain dense for at least a time period of *safe interval*. After that, the state of the region may or may not change.

Next, we proceed to discuss how to build a Quad-tree and compute the safe intervals, followed by how to answer continuous density queries using the Quad-tree.

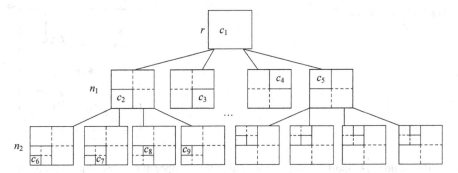

Fig. 6.10 An example of the Quad-tree

6.5.2 Building the Quad-Tree

To facilitate searching dense regions, we partition the space into a grid by employing a Quad-tree. More specifically, the space is recursively divided into four quadrants until the area of the subspace is less than the threshold s given in the density query definition. We set s as the stop condition since it is the minimum area we should consider for a dense region according to the definition. Given a space with an area of S, the depth of the Quad-tree is:

$$L = \lceil \log_4 S/s \rceil + 1 \qquad (6.10)$$

In the Quad-tree, each node corresponds to a cell in the grid. Recall that a node is represented by $((row, col), level, state, safe\_interval)$. The cell can be easily determined by some of these parameters. More specifically, the left-bottom point of the cell is given by:

$$\frac{\sqrt{S}}{2^{level}} \times [row - 1, col - 1] \qquad (6.11)$$

The right-upper point of the cell is given by:

$$\frac{\sqrt{S}}{2^{level}} \times [row, col] \qquad (6.12)$$

Figure 6.10 shows an example of the Quad-tree. Given $S = 32, s = 2$, and $\rho = 1.5$, based on Eq. (6.10), the level number of the Quad-tree is 3. The root of the Quad-tree corresponds to the largest cell c_1. Its level number is 0, the row value is 1, and the col value is also 1. Each internal node is one quadrant of the root, including c_2, c_3, c_4, and c_5. The leaf nodes correspond to the minimum cells (called *leaf cells* hereafter), such as c_6, c_7, c_8, and c_9.

Fig. 6.11 An example of
dense region

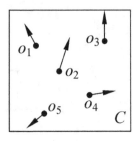

Based on the Quad-tree, initially we count the number of moving objects for each leaf cell and determine if the cell is dense or sparse. By definition, a high-level cell is dense if and only if all the leaf cells below it are dense. For example, in Fig. 6.10, if c_6 through c_9 are dense while some other leaf cell is sparse, then c_2 is returned as a dense region but c_1 is not.

6.5.3   Safe Interval Computation

A safe interval of a dense (sparse) cell means the minimum time period for which the cell is still dense (sparse). Due to the movement of objects, a dense cell may turn into a sparse one, and vice versa. Thus, to support continuous density queries, we maintain the safe intervals for leaf cells of both types, but the safe intervals for high-level cells only if they are dense (i.e., only for dense regions). In the following, we discuss how to compute the safe intervals for dense and sparse leaf cells. The safe interval of a dense high-level cell can be recursively set as the smallest one of its child nodes.

6.5.3.1   Safe Interval of Dense Leaf Cell

For a dense leaf cell, to simplify the computation, we only focus on the objects leaving from it, without considering the entering objects. This is because an entering object will not change the state of a dense cell. It can only change the state of a sparse cell, that is, make the sparse cell dense. Thus, we compute the shortest time interval for which the cell remains dense.

Figure 6.11 shows an example where cell C is dense. There are totally five objects in C, i.e., o_1, o_2, o_3, o_4, and o_5. Let the object number threshold for a dense cell be 3. We compute the time before each object will leave this cell to obtain the *safe interval* of the dense cell. Suppose the leaving times of these objects are t_5, t_3, t_1, t_4, and t_2, sorted in an ascending order. Then, t_1 is the *safe interval* of the dense cell since this cell may become sparse after o_1 leaves.

Algorithm 16 formally describes how to compute the safe interval for a dense leaf *cell*, where $(xmin, ymin)$ and $(xmax, ymax)$ are the bounding coordinates of *cell*, (x, y) is the coordinate of *obj* at time t, and (vx, vy) is the object's speed in the x and y dimensions. We use a heap H to store the last several objects leaving from the cell. Let S_{cell} be the area of the cell. The size of H is set to $\rho \cdot S_{cell}$, which is the density threshold of the cell in terms of the number of objects. For every object in the cell, we compute its leaving time and push the time into H. After processing all the objects, when the object with the minimum leaving time in H leaves from the cell, the object number in the cell will be lower than the density threshold if not considering the objects entering this cell from the outside. Hence, the minimum value in H is the earliest possible time at which that the cell changes its state. This value is returned as the safe interval of *cell*.

Algorithm 16: SIofDense(*cell*)

---

input : The region that needs to be processed
output: Safe Interval of a dense region $cell$
H is a min-heap, whose size is $\rho \cdot S_{cell}$;
for *every obj in cell* **do**
   if *(obj.vx >0)* **then**
      | $lx = cell.xmax - obj.x$;
   else if *(obj.vx <0)* **then**
      | $lx = cell.xmin - obj.x$;
   else
      | $lx = cell.xmax - cell.xmin$;
   end
   if *(obj.vy >0)* **then**
      | $ly = cell.ymax - obj.y$;
   else if *(obj.vy <0)* **then**
      | $ly = cell.ymin - obj.y$;
   else
      | $ly = cell.ymax - cell.ymin$;
   end
   Push $\min(lx/vx, ly/vy)$ into H;
end
Return the minimum value in H;

Note that the safe interval of a dense leaf cell we compute is the shortest time interval for which the dense state remains. Hence, when the safe interval expires, the state of the cell may not be changed if there have been some other objects entering this cell. Thus, the state of this cell and the corresponding safe interval need to be re-calculated upon expiration.

6.5.3.2 Safe Interval of Sparse Leaf Cell

Similar to the dense leaf cell, we only focus on the objects entering the sparse cell, without considering the leaving objects. Suppose that N is the density threshold for the sparse cell and that presently there are M objects in the cell. Then, after $(N - M)$

objects move into this cell, its state might be changed. To reduce the cost of scanning outside objects, we expand the cell level by level until the expanding region contains $(N - M)$ objects. When all the objects in this expanding region enter the cell, the cell's state may be changed. On the other hand, a fast-moving object outside this expanding region may have also entered into the cell. Such earliest time is given by

$$t_o = \frac{L}{V_{max}} \tag{6.13}$$

where V_{max} is the known maximum moving speed and L is the length of the expanding distance. Thus, within the interval t_o, we only need to scan the objects in the expanding region and estimate whether these objects can change the state of this sparse cell by computing their entering times.

Algorithm 17 describes how to compute the safe interval for a sparse leaf cell *cell*. Again, we use a heap H to store the first several objects that will enter to *cell*. The size of H is $(N - M)$. The cell is expanded to a larger region denoted as *Cell* that includes at least $\rho \cdot S_{cell}$ objects. We then compute the entering times of these

---

Algorithm 17: SIofSparse(*cell*)

---

input : The region that needs to be processed
output: Safe Interval of a sparse region *cell*
H is a max-heap, whose size is $(\rho \cdot S_{cell})$−(number of objects in *cell*);
Expand *cell* to *Cell*, which includes at least $(\rho \cdot S_{cell})$ objects;
L is the expanded distance and V_{max} is the maximum velocity of all the objects;
for *every additional object obj in Cell* **do**

    if *(obj.vx > 0 and obj.x ≤ cell.xmin)* **then**
        | $lx = cell.xmin - obj.x$;
    else if *(obj.vx < 0 and obj.x ≥ cell.xmax)* **then**
        | $lx = cell.xmax - obj.x$;
    else
        | $lx = L$;
    end
    if *(obj.vy >0 and obj.y ≤ cell.ymin)* **then**
        | $ly = cell.ymin - obj.y$;
    else if *(obj.vy <0 and obj.y ≥ cell.ymax)* **then**
        | $ly = cell.ymax - obj.y$;
    else
        | $ly = L$;
    end
    $t = lx/vx$;
    if *(obj is not in* cell *at time t)* **then**
        | $t = ly/vy$;
    if *(t > L/V_{max})* **then**
        | $t = L/V_{max}$;
    Push t into H;
end
Return the maximum value in H;

Fig. 6.12 An example of
sparse region

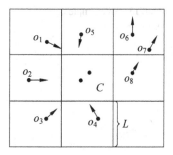

additional objects in *Cell*. If object i's entering time, denoted by t_i, is longer than t_o,
given in Eq. (6.13), t_i is set to t_o. After processing all the additional objects in *Cell*,
the maximum value in H is returned as the safe interval of *cell*.

Figure 6.12 shows an example where C is a sparse region. In the expanding
region, the objects o_1, o_2, o_3, o_4, and o_5 are moving towards C. Suppose that their
entering times are t_2, t_1, t_5, t_3, t_4, sorted in descending order, and that they are all
smaller than t_o. If the region would change to a dense one after three objects move
into it, we will then use t_5 as its safe interval.

Similar to the case for a dense leaf cell, the state and the safe interval of a sparse
cell have to be recomputed when the safe interval expires.

There are two cases in which we need to update the safe interval of a dense/sparse
leaf cell: (1) When the safe interval expires, we need to recompute the state and safe
interval of the cell, as discussed in the last two subsections; (2) when the velocity of
the object changes, we need to recompute the states and safe intervals of those cells
affected by this update. Next we discuss how to deal with the second case.

When the updating object is in a sparse cell, we do not need to recompute
the safe interval of this cell since we consider only the objects entering from the
outside. However, the object may affect the safe intervals of other sparse cells that
the object's moving trajectories cross. We only need to recompute the sparse cells
that the object's new trajectory crosses. For those sparse cells intersected by the old
trajectory, we do not need to recompute their safe intervals until they expire because
before the current safe intervals their states would remain unchanged.

When the updating object is in a dense cell, the safe interval of this cell may be
changed because we compute the safe interval for a dense cell based on the objects
inside the cell. The sparse cells that intersect with the object's new trajectory also
need to be recomputed.

Figure 6.13 shows an example for finding the sparse cells whose safe intervals
need to be recomputed, where S_1, S_2, S_3, S_4, S_5 are sparse cells, D_1, D_2, D_3, D_4
are dense cells, and o_1 is an updating object with its velocity changed. We need not
consider the sparse cells in its old moving direction, i.e., S_3. In the new moving
direction, we identify the sparse cells that o_1 may affect its safe interval. In order to
reduce the computing cost, the formula

$$L_u = v_1 \cdot SI_{max} \tag{6.14}$$

Fig. 6.13 An example of object updating

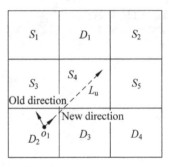

can be used to determine the length of the trajectory, where v_1 is the new speed of o_1 and SI_{max} is the maximum safe interval among all cells. We only update the safe intervals of the sparse cells that intersect with the segment L_u (e.g., S_4 in Fig. 6.13).

6.5.4   Query Processing

Having computed the states and safe intervals for all leaf cells, we now have the required data to identify the dense regions. We search the Quad-tree in a bottom-up manner. For an intermediate node, if all its child nodes are dense (i.e., with the state value of 1), this node is also dense, otherwise it is not, by definition. The bottom-up search of a dense region stops until an ancestor is no longer dense. Then, its child nodes that are dense are returned as answers. The safe interval of the dense region is set as the smallest interval of the leaf cells contained in the dense region. When the safe interval expires, this means the safe interval of a leaf cell expires. The state and safe interval of that leaf cell will be updated, based on which the dense region is also reevaluated. The formal procedure is described in Algorithm 18.

6.6   Summary

In this chapter, we introduced the advanced querying for moving objects including similar trajectory queries and density queries. The cluster-based preprocessing can efficiently support density queries in road networks, and the Quad-tree-based scheme with the notion of safe interval can monitor continuous density queries for moving objects.

Algorithm 18: Query(t)

---

input : Query time t
output: Dense region
for *every leaf node n* **do**
    if *(n.safe_interval $\geq t$)* **then**
        if *(n.state $==$ 0)* **then**
            | break;
        else
            $n'=n$;
            while *(n'.parent.state $==$ 1 and n'.parent.safe_interval $\geq$ query time)*
            $n'=n'$.parent;
            do
                | output n';
            end
            ignore the children of n' and get the next leaf node;
        end
    else
        count number of moving objects in n;
        if *(number$\geq \rho \cdot S_{cell}$)* **then**
            | SIofDense(n);
        else
            | SIofSparse(n);
        end
        if *(n.state changes)* **then**
            adjust value of *n.parent* and take n as the next node;
    end
end

References

1. Agrawal R, Faloutsos C, Swami AN (1993) Efficient similarity search in sequence databases. In: Proceedings of the 4th international conference on foundations of data organization and algorithms (FODO 1993), Chicago, pp 69–84
2. Agrawal R, Lin KI, Sawhney HS, Shim K (1995) Fast similarity search in the presence of noise, scaling, and translation in time-series databases. In: Proceedings of the 21st international conference on very large data bases (VLDB 1995), Zurich, pp 490–501
3. Andoni A, Deza M, Gupta A, Indyk P, Raskhodnikova S (2003) Lower bounds for embedding edit distance into normed spaces. In: Proceedings of the 14th annual ACM-SIAM symposium on discrete algorithms (SODA 2003), Baltimore, pp 523–526
4. Cai Y, Ng R (2004) Indexing spatio-temporal trajectories with chebyshev polynomials. In: Proceedings of the 2004 ACM SIGMOD international conference on management of data (SIGMOD 2004), Paris, pp 599–610
5. Chakrabarti K, Keogh E, Mehrotra S, Pazzani M (2002) Locally adaptive dimensionality reduction for indexing large time series databases. ACM Trans Database Syst 27(2):151–162
6. Chan KP, Fu AW (1999) Efficient time series matching by wavelets. In: Proceedings of the 15th international conference on data engineering (ICDE 1999), Sydney, p 126
7. Chen L, Ng R (2004) On the marriage of edit distance and Lp norms. In: Proceedings of the 30th international conference on very large data bases (VLDB 2004), Toronto, pp 792–803

8. Chen S, Kashyap RL (2001) A spatio-temporal semantic model for multimedia presentations and multimedia database systems. IEEE Trans Knowl Data Eng 13(4):607–622

9. Chon H, Agrawal D, Abbadi AE (2002) Query processing for moving objects with space-time grid storage model. In: Proceedings of the 3rd international conference on mobile data management (MDM 2002), Singapore, pp 121–129

10. Cormode G, Muthukrishnan S (2002) The string edit distance matching problem with moves. In: Proceedings of the 13th annual ACM-SIAM symposium on discrete algorithms (SODA 2002), San Francisco, pp 667–676

11. Douias D, Peucker T (1973) Algorithm for the reduction of the number of points required to represent a line or its character. Am Cartogr 10(42):112–123

12. Ester M, Kriegel HP, Sander J, Xu X (1996) A density-based algorithm for discovering clusters in large spatial databases with noise. In: Proceedings of the 2th international conference on knowledge discovery and data mining (SIGKDD 1996), Portland, pp 226–231

13. Faloutsos C, Ranganathan M, Manolopoulos Y (1994) Fast subsequence matching in time-series databases. In: Proceedings of the 1994 ACM SIGMOD international conference on management of data (SIGMOD 1994), Minneapolis, pp 419–429

14. Hadjieleftheriou M, Kollios G, Gunopulos D, Tsotras VJ (2003) On-line discovery of dense areas in spatio-temporal databases. In: Proceedings of the 8th international symposium on advances in spatial and temporal databases (SSTD 2003), Santorini Island, pp 306–324

15. Jagadish HV, Mendelzon AO, Milo T (1995) Similarity-based queries. In: Proceedings of the 14th ACM SIGACT-SIGMOD-SIGART symposium on principles of database systems (PODS 1995), San Jose, pp 36–45

16. Jensen CS, Lin D, Ooi BC, Zhang R (2006) Effective density queries on continuously moving objects. In: Proceedings of the 22nd international conference on data engineering (ICDE 2006), Atlanta, p 71

17. Keogh E (2002) Exact indexing of dynamic time warping. In: Proceedings of the 28th international conference on very large data bases (VLDB 2002), Hong Kong, pp 406–417

18. Keogh E, Chakrabarti K, Pazzani M, Mehrotra S (2001) Dimensionality reduction for fast similarity search in large time series databases. J Knowl Inf Syst 3(3):263–286

19. Kollios G, Tsotras VJ, Gunopulos D, Delis A, Hadjieleftheriou M (2001) Indexing animated objects using spatiotemporal access methods. IEEE Trans Knowl Data Eng 13(5):758–777

20. Korn F, Jagadish H, Faloutsos C (1997) Efficiently supporting Ad hoc queries in large datasets of time sequences. In: Proceedings of the 1997 ACM SIGMOD international conference on management of data (SIGMOD 1997), Tucson, pp 289–300

21. Martina N and By RA (2004) Spatiotemporal compression techniques for moving point objects. In: Proceedings of the 9th international conference on extending database technology (EDBT 2004), pp 765–782

22. Papadias D, Zhang J, Mamoulis N, Tao Y (2003) Query processing in spatial network databases. In: Proceedings of the 29th international conference on very large data bases (VLDB 2003), Berlin, pp 802–813

23. Pfoser D, Jensen CS, Theodoridis Y (2000) Novel approaches in query processing for moving object trajectories. In: Proceedings of the 26th international conference on very large data bases (VLDB 2000), Cairo, pp 395–406

24. Priyantha N, Miu A, Balakrishnan H, Teller S (2001) The cricket compass for context-aware mobile applications. In: Proceedings of the 7th annual international conference on mobile computing and networking (MOBICOM 2001), Rome, pp 1–14

25. Rafiei D (1999) On similarity-based queries for time series data. In: Proceedings of the 15th international conference on data engineering (ICDE 1999), Sydney, pp 410–417

26. Saltenis S, Jensen CS, Leutenegger ST, and Lopez MA (2000) Indexing the positions of continuously moving objects. In: Proceedings of the 2000 ACM SIGMOD international conference on management of data (SIGMOD 2000), Dallas, pp 331–342

27. Vlachos V, Kollios G, Gunopulos D (2002) Discovering similar multidimensional trajectories. In: Proceedings of the 18th international conference on data engineering (ICDE 2002), San Jose, p 673

28. Yi B, Faloutsos C (2000) Fast time sequence indexing for arbitrary Lp norms. In: Proceedings of the 26th international conference on very large data bases (VLDB 2000), Cairo, pp 385–394

29. Yi B, Jagadish H, Faloutsos C (1998) Efficient retrieval of similar time sequences under time warping. In: Proceedings of the 14th international conference on data engineering (ICDE 1998), Orlando, pp 201–208

Chapter 7
Trajectory Prediction of Moving Objects

Abstract The trajectory prediction is an important part for the management of moving objects. For example, it can be used to improve the performance of the location update strategy and to support the predictive index and queries. In this chapter, we first review some linear prediction methods and analyze their problems in handling moving objects in spatial networks and then present our simulation-based prediction methods: Fast-Slow Bounds Prediction and Time-Segment Prediction. In addition, we also present our uncertain path prediction method.

Keywords Trajectory prediction • Linear prediction • Simulation-based prediction • Uncertain path prediction • Uncertain trajectory mining • Moving object databases

7.1 Introduction

There exist a large number of moving objects in a spatial network with their locations continuously changing. In order to get the location of a moving object in the future time, it is necessary to store its location into a central database via GPS. The research issue is how to accurately maintain the location of a large number of moving objects while minimizing the number of updates. The trajectory prediction plays an important role to solve this problem. Most existing studies propose to lower the update frequency by a trajectory prediction method. They usually use the linear prediction which represents objects locations as linear functions of time. However, the assumption of linear movement in traditional prediction methods limits the applicability in a majority of real-life applications especially in traffic networks where vehicles change their velocities frequently. Moreover, other prediction models with nonlinear prediction proposed by Aggarwal et al. [1] using quadratic predictive function and by Tao et al. [6] based on recursive motion functions for objects with unknown motion patterns improve the precision in predicting the location of each object, but they ignore the correlation of adjacent objects and may not reflect accurately the complex and stochastic traffic movement scenario.

In the management of moving objects, the trajectory prediction method is usually used to improve the performance of the location update strategy and to support the predictive index and queries. In this chapter, we first review some linear prediction methods and analyze their problem in handling moving objects in spatial networks and then present our simulation-based prediction methods: Fast-Slow Bounds Prediction and Time-Segment Prediction, which are more accurate than linear prediction methods in predicting future trajectories of moving objects in spatial networks. Finally we present our uncertain path prediction method, which can predict future trajectories based on the uncertain historical trajectories of moving objects in spatial networks.

7.2   Underlying Linear Prediction (LP) Methods

Most current index and query processing approaches use the linear prediction method for its simplicity and capability of approximating any curve of free movement by piecewise linear segments. Suppose the trajectory function for an object between time t_0 and t_1 is

$$\mathbf{X}_t = \mathbf{X}_{t_0} + \mathbf{V}(t - t_0) \quad (t_0 \leq t \leq t_1) \tag{7.1}$$

where $\mathbf{X}_{t_0}$ denotes the position vector of the object at time t_0 and $\mathbf{V}$ denotes the velocity vector of the object, which is assumed to remain fixed between t_0 and t_1.

7.2.1   General Linear Prediction

The general linear prediction method uses the object's current position $\mathbf{X}_{t_0}$ and current velocity $\mathbf{V}$ to predict its position in the near future. When the prediction is deemed inaccurate, that is, its deviation from the actual position is beyond a predefined threshold, we revise the prediction by resetting $\mathbf{X}_{t_0}$ and $\mathbf{V}$. In situations where object's velocity remains largely constant, this method enables us to make future prediction with high precision. However, when objects move with changing velocity, their trajectory functions have to be revised frequently.

7.2.2   Road Segment-Based Linear Prediction

If objects move in a constrained environment such as a transportation network, we can use the road segments of the network to help model the object's movement. In

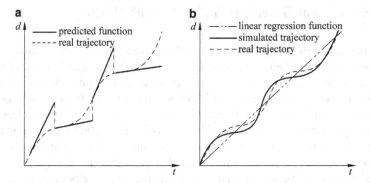

Fig. 7.1 Linear prediction vs. simulation-based prediction. (**a**) General linear prediction. (**b**) Simulation-based prediction

other words, we assume objects move at constant speed along a road segment, that is, their trajectory functions will not change until they move out of a road segment. When an object enters a new road segment, we reset the velocity **V** in its trajectory function. The frequency of revising the trajectory function depends on the average length of the road segments.

7.2.3   Route-Based Linear Prediction

If objects have regular and known routes in the transportation network (e.g., one takes the same route from home to work), we can use the routes instead of the road segments to reduce the number of updates needed to maintain the objects' position. If the route is predicted incorrectly, we simply make an additional update.

However, any real traffic system has a stochastic, dynamic, and fuzzy nature. The accuracy of linear prediction methods mentioned above is inadequate because linear methods can hardly reflect the movement of objects constrained by road networks. For example, in urban road networks, because of traffic conditions, a vehicle may travel at a constant speed, decelerate to stop, wait, accelerate, and travel again at a constant speed. Vehicles may often repeat the above movement in modern urban road networks.

We use Fig. 7.1 to demonstrate the inadequacy of the linear prediction method for real road networks. Figure 7.1a shows the predicted (linear) trajectory and the actual trajectory of an object. We can see that each time the change of the object's velocity is above a certain threshold, an update is triggered and the trajectory is revised by a new velocity vector. The frequent changes of the object's velocity will incur repeated update and prediction.

7.3   Simulation-Based Prediction (SP) Methods

Before presenting the simulation-based prediction methods, we first recall the GCA model introduced in Chap. 2, in particular the definition of CA and the transition of the GCA model. A cellular automaton (CA) consists of a finite oriented sequence of cells. In a configuration, each cell is either empty or contains a symbol. During a transition, symbols can move forward to subsequent cells, symbols can leave the CA, and new symbols can enter the CA. Let i be an object moving along an edge. Let $v(i)$ be its velocity, $x(i)$ its position, $gap(i)$ the number of empty cells ahead (forward gap), and $P_d(i)$ a randomized slowdown rate that specifies the probability that it slows down. We assume that V_{max} is the maximum velocity of the moving objects. At each transition of GCA, each object changes velocity and position in a CA of length L according to the rules below:

1. If $v(i) < V_{max}$ and $v(i) < gap(i)$, then $v(i) \leftarrow v(i) + 1$.
2. If $v(i) > gap(i)$, then $v(i) \leftarrow gap(i)$.
3. If $v(i) > 0$ and $rand() < P_d(i)$, then $v(i) \leftarrow v(i) - 1$.
4. If $(x(i) + v(i)) \leq L$, then $x(i) \leftarrow x(i) + v(i)$.

Considering the simulation feature of the GCA model, we use GCAs not only to model road networks but also to simulate future trajectories of moving objects by the transitions of GCAs, where objects' movement follows traffic rules. Based on the GCA, a *simulation-based prediction (SP)* method to anticipate future trajectories of moving objects is proposed. The SP method treats the object's simulated results as its predicted positions to obtain its future in-edge trajectory. To refine the accuracy, based on different assumptions on the traffic conditions, we simulate two future trajectories in discrete points for each object on its edge. Then, by linear regression and translating, the trajectory bounds that contain all possible future positions of a moving object on that edge can be obtained. When the object moves to another edge in the GCA or the predicted position exceeds its actual position above the predefined accuracy, another simulation and regression will be executed to predict new future trajectory bounds. The process of the simulation-based prediction can be seen in Fig. 7.2.

7.3.1   Fast-Slow Bounds Prediction

Most existing work uses the CA model for traffic flow simulation in which the parameter $P_d(i)$ is treated as a random variable to reflect the stochastic, dynamic nature of traffic system. However, we extend this model for predicting the future trajectories of objects by setting $P_d(i)$ to values that model different traffic conditions. For example, laminar traffic can be simulated with $P_d(i)$ set to 0 or a small value, and the congestion can be simulated with a larger $P_d(i)$. By giving $P_d(i)$ two values, we can derive two future trajectories, which describe,

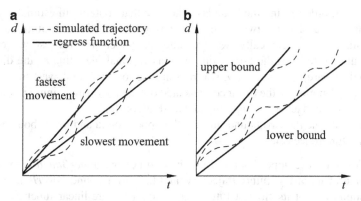

Fig. 7.2 Two predicted bounds of future trajectories. (**a**) Simulated trajectories. (**b**) Two predicted bounds

respectively, the fastest and slowest movements of objects as shown in Fig. 7.2a. In other words, the object's future locations are most probably bounded by these two trajectories. The value of $P_d(i)$ can be obtained by sampling from the given dataset.

For getting the future trajectory function of an object from the simulated discrete points, we need to regress the discrete positions. We find that in most cases, the linear regression (as shown in Fig. 7.2a) fits the prediction well and at low cost. The ordinary least square estimation (OLSE) method, for example, can be calculated efficiently at low data storage cost. Let the discrete simulated points be $(t_1, d_1), \ldots, (t_i, d_i), \ldots, (t_n, d_n)$, where d_i $(i \in [1, n])$ denotes the relative distance in a network edge. The average value of them be $\bar{t}$ and $\bar{d}$. After regression, the trajectory function of the moving object is

$$D(t) = \hat{\beta}_0 + \hat{\beta}_1 \cdot t \tag{7.2}$$

where $\hat{\beta}_o$ and $\hat{\beta}_1$ are given by

$$\hat{\beta}_0 = \bar{d} - \hat{\beta}_1 \cdot \bar{t} \tag{7.3}$$

$$\hat{\beta}_1 = \frac{\sum_{i=1}^{n} t_i d_i - n\bar{t} \cdot \bar{d}}{\sum_{i=1}^{n} t_i^2 - n(\bar{t})^2} \tag{7.4}$$

In Fig. 7.2a, the dashed curves show two future trajectories, which are the slowest and the fastest movements simulated by using different P_d. Applying the OLSE algorithm to the two trajectories generates two linear functions, which are shown in solid lines.

$$fastTrj: \quad D(t) = \alpha_f \cdot t + \gamma_f \tag{7.5}$$

$$slowTrj: \quad D(t) = \alpha_s \cdot t + \gamma_s \tag{7.6}$$

Finally, in order to find the bounds of the area that contains all estimated future positions, we translate the two regression lines, until all estimated future positions fall within. More specifically, we translate the upper line (fastest movement) upward until it touches the point with the max residual (denoting ε_1 the distance translated upward), and similarly, we translate the lower line (slowest movement) downward (denoting ε_2 the distance translated downward). This minimizes the loss of information and errors brought by the OLSE algorithm.

We now define the two bound lines as the upper bound and lower bound of the object's future trajectories.

Definition 7.1. The upper bound of an object trajectory *upperBound* is the upper bound line of its fastest future trajectory, and the lower bound *lowerBound* is the lower bound line of its slowest future trajectory. They are linear functions of the following form:

$$upperBound: \quad D(t) = \alpha_f \cdot t + \lambda_f \tag{7.7}$$

$$lowerBound: \quad D(t) = \alpha_s \cdot t + \lambda_s \tag{7.8}$$

where $\lambda_f = \gamma_f + \varepsilon_1, \lambda_s = \gamma_s - \varepsilon_2$.

The two bound lines are shown in Fig. 7.2b. We can treat the two predicted lines as the bounds of the possible future positions of one object. The predicted trajectory bounds can be used in the predictive index structure and query processing in road network to reduce the index updates and filter unnecessary query results to improve the performance of predictive query. For example, given a predictive range query with the specified region R during time interval $[t_1, t_2]$ in the future, we can filter the objects in the result during the preprocess phase if the area between their upper and lower trajectory bounds cannot intersect the R during $[t_1, t_2]$.

However, for other applications such as the tracking of moving objects, a single predicted function is needed to obtain the specific future positions of the object. For example, to lower update frequency from moving objects to server database, a general principle for location update policies is as follows: the moving objects equipped by GPS receiver do not report their locations to the server unless their actual positions exceed the predicted positions to a certain threshold. Their predicted positions need to be computed by a single predicted function. In this case, we can also adapt the SP method to obtain a compact and simple linear prediction function. The process can be seen in Fig. 7.3. After regressing the two simulated future trajectories to two linear function denoting L_1 and L_2, we compute the middle straight line L_3, the bisector of the angle a between L_1 and L_2 as the final predicted function $L(t)$.

Although the predicted function obtained by the SP method is a simple linear function, it is different from the linear prediction in that the SP method not only considers the speed and direction of each moving object but also takes correlation of objects as well as the stochastic behavior of the traffic into account.

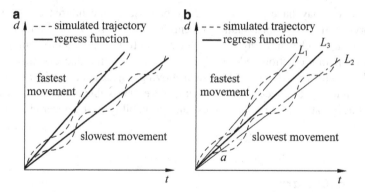

Fig. 7.3 Singe predicted future trajectory. (**a**) Simulated trajectories. (**b**) Single predicted function

7.3.2   Time-Segmented Prediction

As the prediction of in-edge trajectory only uses the GCA to simulate the movement of objects in an edge, we have to consider the cases when objects move across the nodes in order to make the global trajectory prediction. If the out-degree of a node in the GCA is one, the behavior of the object in the adjacent edge is the same. However, if the out-degree of the node is bigger than one, we cannot trace the objects cross different edges. In this case, we could use the probability of objects changing the edges according to the historical data.

In the last section, we only predict the in-edge trajectory of the object moving in one edge of the GCA. When the object moves to another edge or its prediction accuracy of the future positions cannot meet the given accuracy requirement, we issue another prediction based on the current traffic conditions. For the predicted fast and slow trajectory bounds, it is possible that the predicted positions at different time stamps exceed the real positions given query precision range. In particular, as the time goes, the predicted trajectory bounds will expand and lead to worse prediction accuracy. Therefore, the Time-Segmented prediction method is used in this case. The simulation and prediction are issued every fixed time internal, such as *tLength*. Even within the *tLength*, when the predicted locations cannot meet the requirement of query anymore, we issue another prediction. The Time-Segmented prediction method can estimate the real trajectory of moving objects with better accuracy.

7.4   Uncertain Path Prediction Methods

Most of the moving object path prediction methods proposed so far assume that the input historical trajectory is a complete path. However, in reality, as the moving objects periodically update their location to server, historical trajectory reconstruction may have great uncertainty because of the inference on discrete trajectory points. Such uncertainty prevents the practical use of traditional path prediction approaches in many LBS applications. To handle this problem, a novel method for path prediction under network constraints is introduced in this section, particularly with uncertainty of historical moving object path considered. It uses trajectory interpolation technique to generate several possible paths and then mines out the future path based on the hotness and probability of the uncertain paths for prediction.

7.4.1   Preliminary

As historical path could be incomplete due to periodical location update, it is necessary to derive the complete historical path using interpolation techniques. In this section, we define and explain some useful notions first.

Given two sequential location updates of a moving object, suppose that they are mapped to road segments s and e, respectively, using map-matching algorithms, and s and e are not adjacent in the road network. Finding the actual path between s and e, the moving object actually passed should find out all the possible paths and then verify them.

Definition 7.2. Suppose s and e are two road segments mapped by two continuous location updates of a moving object, and they are not adjacent in the road network, and there exists m possible paths between s and e, which are $R_1, R_2, R_3, \ldots, R_m$, then the *Path Probability* of $P_{R_i|e}$ is defined as:

$$
P_{R_i|s} = \begin{cases} \dfrac{\sum\limits_{k=1}^{m} L(R_k) - L(R_i)}{(m-1) \sum\limits_{k=1}^{m} L(R_k)} & (m > 1) \\ 1 & (m = 1) \end{cases}
\tag{7.9}
$$

where $L(R_i)$ is the length of the path R_i and m is the number of possible paths between s and e.

Definition 7.3. The combination of a path and its corresponding path probability is defined as an *uncertain item*, which is represented as $r : p$, where r presents the path that a moving object may pass and p presents the probability of passing path r.

Definition 7.4. The *uncertain itemset* is a dataset composed of a series of uncertain items, which is represented as $I = (r_1 : p_1, r_2 : p_2, \ldots, r_m : p_m)$.

Definition 7.5. The *uncertain path sequence* is an ordered sequence of a series of uncertain itemsets, which is represented as $S = < I_1, I_2, \ldots, I_n >$.

Figure 7.4 is an example of an uncertain path sequence. As shown in this figure, location updates occur in road segment s, f, and e; the dotted lines indicate the

Fig. 7.4 Uncertain path sequence

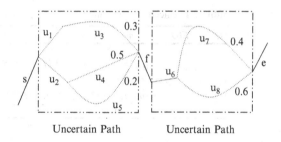

Uncertain Path   Uncertain Path

Fig. 7.5 Uncertain path prefix tree

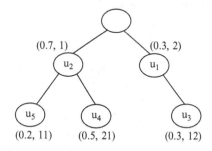

possible paths from s to f and from f to e, respectively. The uncertain path sequence S is composed of five uncertain itemsets, and it can be represented as $S =< s : 1, (u_1u_3 : 0.3, u_2u_4 : 0.5, u_2u_5 : 0.2), f : 1, (u_6u_7 : 0.4, u_6u_8 : 0.6), e : 1 >$.

The movement of most objects usually follows a periodical pattern, for example, most people would get up at the same time and then choose identical or similar way to work every day. Therefore, it is essential to discover the periodical trajectory patterns from the historical trajectory and then make use of them.

A data structure called uncertain path prefix tree is used for the frequent trajectory pattern mining. Each uncertain itemset in an uncertain path sequence is mapped to an uncertain path prefix tree: as shown in Fig. 7.4, uncertain itemsets $I_2 = (u_1u_3 : 0.3, u_2, u_4 : 0.5, u_2u_5 : 0.2)$ and $I_4 = (u_6u_7 : 0.4, u_6u_8 : 0.6)$ are mapped to the two uncertain path prefix trees, respectively. Based on the road segments that appear in itemsets, we further construct the uncertain path prefix tree structure according to their partial order in all possible paths.

In the uncertain path prefix tree, each node contains a two tuple of (RID, P), where RID is the ID of a path and P is path probability. Note that path probability of a leaf node indicates the possibility from the root node to this node, while P of internal node is the sum of probability weight of all its child nodes. Both height and RID of root node are 0. Given a node w in the h height level, assume it is the ith child node of node v; the RID of node w is represented as $RID(w) = i \times 10^{h-1} + RID(v)$. Figure 7.5 shows the structure of the uncertain path prefix tree of I_2. According to Property 7.1, we can judge whether two nodes belong to a same path according to their RID.

Table 7.1 Uncertain path dataset

SID	EID	(Item, RID, P)
1	1	(s, 0, 1)
1	2	(u1, 2, 0.3) (u2, 1, 0.7) (u3, 12, 0.3)(u4, 21, 0.5) (u5,11,0.2)
1	3	(f, 0, 1)
1	4	(u6, 1, 1) (u7, 21, 0.4) (u8, 11, 0.6)
1	5	(e, 0, 1)

Table 7.2 id-list sample

Item	SID	EID	RID	P
s	1	1	0	1
u_1	1	2	2	0.3
u_2	1	2	1	0.7
u_3	1	2	12	0.3
u_4	1	2	21	0.5
u_5	1	2	11	0.2
f	1	3	0	1
u_6	1	4	1	1
u_7	1	4	21	0.4
u_8	1	4	11	0.6
e	1	5	0	1

Property 7.1. Suppose that RID_v and RID_w are the path number of nodes v and w in an uncertain path prefix tree, v is the ancestor of w if $(RID_w > RID_v) \cap ((RID_w - RID_v)\&RID_v = 0)$ can be satisfied.

In Fig. 7.5, we can derive that $RID(u_2) = 1$, $RID(u_5) = 11$, $RID(u_3) = 12$. According to Property 7.1, road segment u_2 is the ancestor of road segment u_5. In other words, they belong to the same path and u_2 is in order before u_5. In contrast, it can be inferred that u_2 and u_3 do not belong to a same path.

In the uncertain path dataset D, each uncertain path sequence has a unique identification number SID; each uncertain itemset also has a unique identification number EID. According to uncertain path prefix tree, each uncertain itemset can be convert to a 3-tuple $(Item, RID, P)$, where $Item$ is the road segment. Uncertain path dataset D can be represented as Table 7.1.

7.4.2 Uncertain Trajectory Pattern Mining Algorithm

The goal of uncertain trajectory mining is to mine out frequent trajectory patterns from uncertain path dataset D. The support of a frequent trajectory pattern is the sum of path probability in dataset D. A data structure named id-list similar to Zaki [8] is used in the algorithm, where each road segment has an id-list as shown in Table 7.2.

Table 7.3 id-list of road segment sequence su_7

su_7							
SID	EID(s)	EID(u_7)	RID(s)	RID(u_7)	P(s)	P(u_7)	P(su_7)
1	1	4	0	21	1	0.4	0.4

The basic idea of uncertain trajectory pattern mining algorithm is that the set of frequent 1-itemsets is found firstly, then this set is used to find the set of frequent 2-itemsets, which is further used to find the set of frequent 3-itemsets, and so on, until no more frequent k-itemsets can be found. It appears similar to Apriori, but they use completely different ways to compute the support value and generate candidate set, toward which the computational process can be finished with the help of id-list. Table 7.3 shows the id-list of road su_7.

Property 7.2. Given two frequent trajectory patterns Ac and Ad (same as the patterns in Apriori), suppose their path probability are P_1 and P_2, respectively, then we can compute the path probability of path Acd as (1) If $EID(c) > EID(d)$, then $P = 0$. (2) If $EID(c) < EID(d)$, then $P = P_1 \times P(d)$. (3) If $EID(c) = EID(d)$, then $P = P_2$.

Property 7.2 provides a way to compute path probability based on its partial paths, and the support of trajectory pattern can be derived accordingly (i.e., its path probability). If $EID(c) > EID(d)$, it means c is ancestor of d, so the path probabilistic of Acd is 0. If $EID(c) < EID(d)$, we know that c and d exist in different uncertain itemsets, so the path probability of Acd is $P = P_1 \times P(d)$. Otherwise, if $EID(c) = EID(d)$, c and d exist in the same uncertain itemset, and as a result the path probability of Acd is $P = P2$.

Property 7.3. Candidate trajectory pattern Acd can be generated by Ac and Bd as follows: A and B must have the same value of *SID, EID, RID* according to their uncertain item and satisfy at least one of the following two conditions: (1) $EID(c) < EID(d)$ and $EID(d) - EID(c) \leq WIDTH$ (2) $EID(c) = EID(d)$ and $(RID_d > RID_c) \cap ((RID_d - RID_c)\&RID_c = 0)$

Property 7.3 is the rule for candidate itemsets generation. When $EID(c) < EID(d)$, the difference between $EID(d)$ and $EID(c)$ should less than a predefined threshold *WIDTH*, so as to ensure the effectiveness of prediction. When $EID(c) = EID(d)$, it is obvious that Acd is a valid candidate only if c is the ancestor of d. Details of uncertain trajectory pattern mining are shown in Algorithm 19.

7.4.3   Frequent Path Tree

To facilitate path estimation, we construct a data structure called Frequent Path Tree (FPTree) for future path search. FPTree is a variation of Signature Tree [2, 4].

Algorithm 19: Uncertain trajectory pattern mining

---
input : Minimum Support: *MIN_SUP*, Uncertain Path Data Set: *D*, Threshold: *WIDTH*
*F*1 = the set of frequent 1-itemsets;
*F*2 = the set of frequent 2-itemsets;
for $k = 3; F_{k-1} \neq \emptyset; k + + $ **do**
   | F_k = Enumerate(F_{k-1}, *MIN_SUP*, *WIDTH*);
end
procedure Enumerate(F_{k-1}, *MIN_SUP*, *WIDTH*) $T = \emptyset$;
for $\forall A_i \in F_{k-1}$ **do**
   | **for** $\forall A_j \in F_{k-1}$ **do**
   | | $R = A_i$ APPEND A_j;
   | | **if** *(R is frequent)* **then**
   | | | $T = T \cup R$;

   | **end**
end
return T;

Similar to [5], each node of FPTree contains entries of the form $< sig, ptr >$. In a leaf node entry, *sig* is the signature of a transaction and *ptr* is a transaction id. Each internal node entry is the logical OR on all signatures in its subtree. However, there are two shortcomings in Signature Tree: (1) It has high storage space cost because of the long bitmap of signature tree, and (2) it also could be computationally expensive. Therefore, we use hierarchical bitmap to compress the storage of bitmap in Signature Tree for complexity reduction purpose.

Given that the bitmap is usually sparse, great storage space can be saved if we adopt hierarchical tree structure to compress it. In a hierarchical bitmap, each node is formed by a l-bit bitmap and l pointers to its child nodes. Similar to Quadtree, it is based on a divide-and-concur method: the big bitmap is divided into l parts from the root level. If all bits in the kth part of the bitmap are zero, then the kth bit is simply set as zero and the kth pointer is set as null. Otherwise, if not all of them are zero, the kth bit is set as 1, and we further partition the kth part of bitmap in the same way. This process continues until the length of node bitmap is l.

Figure 7.6 is an example of a hierarchical bitmap set $S = 2, 3, 9, 12, 61$ where $l = 4$. The nodes with dashed frame are empty nodes, i.e., all bits in this node are zero. The hierarchical bitmap tree only index non-null nodes. As shown in this graph, even though the hierarchical bitmap tree added 3 additional non-leaf index nodes, it actually saved 13 leaf nodes' storage space. That means, the space compression rate is around 66%.

The leaf node in FPTree contains entries in the format of $<key, c, ptr>$, where *key* is a pointer point to the hierarchical bitmap, c is the support of trajectory

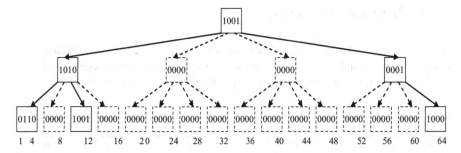

Fig. 7.6 Hierarchical bitmap

Trajectory Pattern	Support
2,3,9,12 ,61	0.6
2,3,25,60	0.2
13,16	0.5
13,15	0.4

Fig. 7.7 Trajectory pattern and support

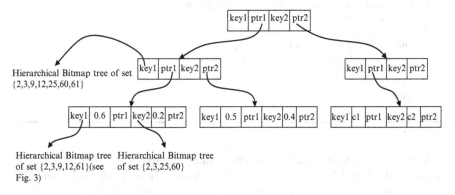

Fig. 7.8 Frequent path tree

pattern, and *ptr* is a pointer points to the trajectory pattern. The internal node in FPTree contains entries of the form *<key, ptr>*; *key* is a pointer to the hierarchical bitmap of the logical OR on all bitmaps in its subtree. *ptr* is a pointer to the child node.

Figure 7.7 shows a set of trajectory patterns and their support, e.g., the first pattern is the trajectory passing road segments (2, 3, 9, 12, 61) in a sequential order. Figure 7.8 shows its corresponding FPTree.

7.4.4   Trajectory Prediction

After searching all similar trajectory patterns from FPTree, each similarity of
trajectory pattern and query path is computed. The most similar trajectory pattern is
used for prediction; the path prediction algorithm is shown as Algorithm 20.

Algorithm 20: Path prediction

---

   input : Query Trajectory:Q, Set of Trajectory Pattern:$P = \{P_1, P_2, \ldots, P_n\}$, Support of
            Trajectory Pattern:$Sup = \{s_1, s_2, \ldots, s_n\}$
   $S = \varnothing$;
   $P\_Max = \varnothing$;
   for $\forall P_i \in P$ **do**
      | $s_i = $ Similarity(P_i, Q);
      | $S = S \cup \{s_i\}$;
   end
   for $\forall s_i \in S$ **do**
      | **if** *(s_i is most similar)* **then**
      | | $P\_Max = P\_Max \cup \{P_i\}$;

   end
   if *(there is only one trajectory pattern in P_Max)* **then**
      | **return** P_Max;
   else
      | **for** $\forall p\_maxi \in P\_Max$ **do**
      | | **if** *($p\_maxi\,has\,the\,largest\,support$)* **then**
      | | | **return** p_maxi;

      | **end**
   end

In Algorithm 20, the trajectory patterns similar to a given (uncertain) query path
are selected out by FPT-index and saved in set S first. Then the similarity between
trajectory pattern and the query path is computed based on the total weight of their
common road segments. Future path predictions are made based on the ranking of
the similarity values of trajectory patterns to the given query path.

7.5   Other Nonlinear Prediction Methods

The prediction model plays an important role in tracking of moving objects. Most
existing prediction methods assume linear movement, which limits applicability in
the majority of real applications. In paper [1], the nonlinear models such as the
acceleration are used to represent the trajectory which is affected by the abnormal
traffic such as traffic incident. Xu and Wolfson [7] apply the time-series prediction
together with moving speed to traffic management and moving object databases.

Karimi and Liu [3] describe a technique for trajectory prediction which assigns probabilities to the roads emanating from an intersection and uses the most probable route within some extracted sub-road network to predict. Recently, according to the trend of each object's own movement regarding its recent past locations, Tao et al. [6] propose a prediction method based on recursive motion functions for objects with unknown motion patterns. Although these prediction methods can improve the precision of location prediction of each object, they ignore the correlation of movements of adjacent objects in traffic networks, and thus may not reflect the realistic traffic movements.

7.6 Summary

Some trajectory prediction methods are introduced in this chapter, which are very important to the management of moving objects. Motivated by the features of vehicle's movements in traffic networks, we propose the new simulation-based prediction methods, which are much more precise than the linear prediction methods. In Chaps. 3 and 4, we have used the simulation-based method to improve the performance of the location update strategy and to support the predictive index and queries. In addition, we also propose the new uncertain path prediction method, which can predict future path based on uncertain historical trajectories.

References

1. Aggarwal C, Agrawal D (2003) On nearest neighbor indexing of nonlinear trajectories. In: Proceedings of the 22nd ACM SIGMOD-SIGACT-SIGART symposium on principles of database systems (PODS 2003), San Diego, pp 252–259
2. Hellerstein JM (1994) The Rd-tree: an index structure for sets. Technical report no. 1252, University of Wisconsin, Madison
3. Karimi HA, Liu X (2003) A predictive location model for location-based services. In: Proceedings of the 11st ACM international symposium on advances in geographic information systems (GIS 2003), New Orleans, pp 126–133
4. Mamoulis N, Cheung D, Lian W (2003) Similarity search in sets and categorical data using the signature tree. In: Proceedings of the 19th international conference on data engineering, Bangalore, 2003. IEEE Computer Society, Washington, DC, pp 75–86
5. Morzy M, Morzy T, Nanopoulos A (2003). Hierachical bitmap index: an efficient and scalable indexing technique for set-valued attribute. In: Proceedings of ADBIS'03, Berlin, pp. 236–252
6. Tao Y, Faloutsos C, Papadias D, Liu B (2004) Prediction and indexing of moving objects with unknown motion patterns. In: Proceedings of the 2004 ACM SIGMOD international conference on management of data (SIGMOD 2004), Paris, pp 611–622
7. Xu B, Wolfson O (2003) Time-series prediction with applications to traffic and moving objects databases. In: Proceedings of the 3rd ACM international workshop on data engineering for wireless and mobile access (MobiDE 2003), San Diego, pp 56–60
8. Zaki MJ (2001) SPADE: an efficient algorithm for mining frequent sequences. Mach Learn 42(1/2):31–60

Chapter 8
Uncertainty Management in Moving Objects Database

Abstract The uncertainty is mainly caused by measurement error and sampling error, which makes uncertainty as an inherent aspect of moving object database. To manage uncertainty, lots of research has been proposed with lots of effective models and algorithms. This chapter presents a systematic overview of the various issues and solutions related to the uncertainty management in moving objects database. In the first part of this chapter, we introduce three representative models to illustrate how the uncertainty can be managed in moving object database. Afterwards, a novel modeling framework is presented to manage uncertain trajectory and define some database operations related to the framework.

Keywords Uncertainty management • Uncertainty trajectory • Spatial network • Moving object databases • Abstract data type

8.1 Introduction

Uncertainty management is one of the most important issues in moving objects databases. In the MOD system, moving objects such as cars, flights, ships, and pedestrians are uniquely identified, and each of them is equipped with a portable computing platform and other integrated location tracking equipments like GPS. Through location updates, moving objects report their latest location information to server so that the location of any object at any time can be retrieved by users through query. Since location update is made intermittently, between any two consecutive location updates, the server cannot tell the exact location of this object, so it is important to find ways for inference. As shown in Fig. 8.1, given a query on the location of an object at t_3, the result cannot be directly obtained from the database server. As a result, uncertainty becomes an inherent aspect of MOD [10, 16].

In recent years, lots of research has been focused on the uncertainty management problem, with lots of effective models and algorithms being proposed. Generally speaking, there are two kinds of uncertainty: measurement error and sampling error.

X. Meng et al., *Moving Objects Management: Models, Techniques and Applications*, DOI 10.1007/978-3-642-38276-5_8,

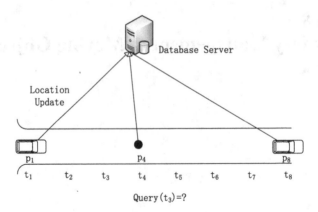

Fig. 8.1 Location uncertainty

With the help of GPS devices, measurement error can be very small compared with the sampling error. As for the sampling error, uncertainty depends directly on the update frequency of moving objects.

In [10], the authors analyze the sources of uncertainty in moving objects databases and proposed a framework to deal with uncertain data. In [15, 16], the authors discussed the uncertainty management strategies in the DOMINO system. By applying an uncertainty threshold, the trajectory of a moving object is extended from a curve to a tube in the $X \times Y \times T$ space, and the operations, such as *inside*, are extended by introducing the uncertainty semantics such as "sometimes," "always," "possibly," and "definitely." Research [12] explored the uncertainty and fuzziness in managing moving objects, and a framework is provided to deal with spatio-temporal indeterminacies. In [14], a set of data types and operations have been proposed for the uncertainty management of moving objects. However, all the above studies are based on the $X \times Y \times T$ Euclidean space, and nearly none of them have treated the interaction between moving objects and the underlying transportation networks in any way.

Recently, based on the fact that most moving objects only move in fixed transportation networks, researchers have realized the importance of modeling network-constrained moving objects, and meanwhile, the uncertainty of network-constrained moving objects has also been studied. The uncertainty management problem for network-constrained moving objects is analyzed in detail in [6]. Through reasonable location modeling and location update methods, the possible location of a moving object at any time is reduced to a graph route section instead of a region so that the indeterminacies can be greatly reduced. In [8], authors propose to use transportation networks to reduce sampling noises from GPS or to predict future positions of moving objects. In [2], authors further discuss the uncertainty of network-constrained moving objects based on the study [6], with a rich set of data types and operations for representing historical uncertain trajectories defined.

Even though the modeling of uncertainty has been relatively well studied, the research on the operations of moving object trajectories with uncertainty considered is very limited. Previously, a lot of index methods were proposed to deal with the trajectories of moving objects both with and without network concerned. For instance, in [11, 13], the authors proposed an R-tree-based index method for Euclidean space-based trajectory data. In [7], the author proposed a Fixed Network R-Tree (FNR-Tree) to index moving objects on fixed networks. In [1], authors proposed the MON-tree to further improve the FNR-Tree. Also, the study in [3] deals with the future trajectories of network-constrained moving objects.

In this chapter, we first present some representative uncertainty management approaches. After that we introduce a novel framework that can manage uncertainty trajectory effectively and answer queries about them accurately. Particularly, we focus on the key technical issues like uncertain trajectory modeling, database operations, and query processing for uncertainty management.

8.2 Representative Models

8.2.1 2D-Ellipse Model

To depict the uncertainty of moving object, Pfoser and Jensen in [10] present a 2D-ellipse model in spatial space. An error ellipse is proposed to measure the uncertainty of moving object according to its maximum speed and the sampling interval. Suppose P_1 and P_2 are two consecutive samples from a moving object, the sampling interval is Δt, and its maximum speed is v_m; then the possible location of the moving object at time t_x, $t_1 < t_x < t_2$, can be computed. If the object moves at v_m from P_1 and its trajectory is a straight line, its position at time t_x will be on a circle of radius $r_1 = v_m(t_1 + t_x)$ around P_1. Thus, the points on the circle represent the furthest away from P_1 the object can reach at time t_x. If the object's speed is lower than v_m, or its trajectory is not a straight line, the object's position at time t_x will be somewhere within the area bounded by the circle of radius r_1. Applying the same assumptions again, the object's position at time t_x is on the circle with radius $r_2 = v_m(t_2 - t_x)$ around P_2. If the object moves slower or its trajectory is not a straight line, it is somewhere within the area bounded by this circle.

As shown in Fig. 8.2, with the evolving sampling error, the possible locations of the moving object in between two consecutive samples lie on an error ellipse with positions P_1 and P_2 as its foci. The length of the semimajor axis is $2a = r_1 + r_2$. As the definition of an ellipse points out, it is a curve consisting of all points in the plane whose sum of distances, r_1 and r_2, from two fixed points, P_1 and P_2 (the foci) separated by a distance of $2c$, is a given constant, $2a$. The measure $2c$ can be interpreted as the observed distance between P_1 and P_2, whereas $2a$ is the maximum distance the object can travel. The "thickness" of the ellipse, $2b$, is determined by the equation $b^2 = a^2 - c^2$. This means that the smaller the difference between

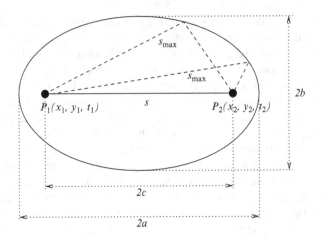

Fig. 8.2 Error ellipse

the observed distance $2c$ and the maximum distance $2a$, the "thinner" the ellipse. In extreme cases, the ellipse degrades to a line segment. In worst cases, where the object does not move between consecutive position samples, the ellipse becomes a circle.

8.2.2 3D-Cylinder Model

Traditionally, the trajectory of a moving object was modeled as a polyline in three-dimensional space (two dimensions for geography, and one for time). In order to capture uncertainty, Trajcevski et al. in [16] propose an approach to model the trajectory as a cylindrical volume in 3D. Traditionally, spatio-temporal range queries ask for the objects that are inside a particular region, during a particular time interval. However, for the moving objects, one may query the objects that are inside the region *sometime* during the time interval, or for those *always* inside during the time interval. Similarly, one may query the objects that are *possibly* inside the region or for the ones that are *definitely* there.

An uncertain trajectory is obtained by associating an uncertainty threshold r with each line segment of the trajectory. For a given motion plan, the line segment together with the uncertainty threshold constitutes an "agreement" between the moving object and the server. The agreement specifies the following: the moving object will update the server if and only if it deviates from its expected location (according to the trajectory) by r or more. How does the moving object compute the deviation at any point in time? Its onboard computer receives a GPS update every 2 s, so it knows its actual location. Also, it has the trajectory, so by interpolation, it can compute its expected location at any point in time. The deviation is simply the distance between the actual and the expected location.

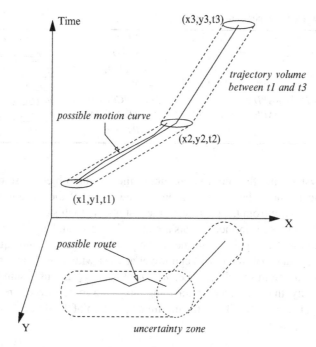

Fig. 8.3 Possible motion curve and trajectory volume

Accordingly, Trajcevski et al. in [16] define the *uncertainty threshold*, *r-uncertainty area*, and *trajectory volume*. *Uncertainty threshold* is defined as a tuple (Tr, r), where Tr represents a trajectory and r is a positive real number, denoting the uncertainty threshold. For each point (x, y, t) on the Tr, its *r-uncertainty area* is a horizontal circle with radius r, centered at (x, y, t), where (x, y) is the expected location at time $t \in [t_1, t_n]$. Considering all the $t \in [t_1, t_n]$ in 3D space, the continuous *r-uncertainty area* constitutes a 3D volume called *trajectory volume* shown in Fig. 8.3.

8.2.3 Model the Uncertainty in Database

We recall the model presented in [9] for moving objects in network that the most important type of moving object is the moving point object. With this abstraction, we can model the movement of cars, trains (if their extent is ignored), people, etc., and implement them in the database system.

Almeida and Güting in [2] point out when an update in the DBMS is triggered by an object, four possibilities can occur: The object is new to the system and starts its movement; the object changes routes; the object position deviation exceeds the threshold ξ; or the object speed deviation exceeds the threshold ψ. If the object is

Table 8.1 Abstract data types with uncertainty

	→ *BASE*	*int,real,string,bool*
	→ *SPATIAL*	*point,points,line,region*
	→ *GRAPH*	*gpoint,gline*
	→ *TIME*	*instant*
$BASE \cup SPATIAL \cup GRAPH$	→ *UNCERTAIN*	*uncertain*
$BASE \cup SPATIAL \cup GRAPH \cup UNCERTAIN$	→ *TEMPORAL*	*moving, intime*
$BASE \cup TIME$	→ *RANGE*	*range*

new to the system, the first information about the moving object is stored in the database. And if under the rest conditions, a new piece of the trajectory will be added to the database from the position in the last update to this current one.

Assume that the measurement points are at positions p_1 and p_2 of the specified route r taken at times t_1 and t_2. At time t_1, the position is precise and equals to p_1 with speed v_1. After that, the object can travel at least with $v_1 - \psi$ and at most at $v_1 + \psi$ or its maximum speed v_{max}, i.e., $min v_1 + \psi, v_{max}$. Let us assume, without loss of generality, that the object is traveling along the route side where positions are increasing, i.e., $p_1 \leq p_2$. Thus, the minimum position of the object at some time $t, t_1 < t < t_2, p_{min}(t)$ is

$$p_{min}(t) = max\{p_1 + max v_1 - \psi, 0(t - t_1), \tag{8.1}$$

$$p_1 + \xi + \frac{p_2 - p_1}{t_2 - t_1}(t - t_1), \tag{8.2}$$

$$p_2 - min\{v_1 + \psi, v_{max}\}(t_2 - t)\} \tag{8.3}$$

Analogously, the maximum position of the object at some time t, $p_{max}(t)$ is

$$p_{max}(t) = min\{p_1 + min\{v_1 + \psi, v_{max}\}(t - t_1), \tag{8.4}$$

$$p_1 + \xi + \frac{p_2 - p_1}{t_2 - t_1}(t - t_1), \tag{8.5}$$

$$p_2 - max\{v_1 - \psi, 0\}(t_2 - t)\} \tag{8.6}$$

These two equations give us the geometry of a moving object between two measurement points p_1 and p_2 at times t_1 and t_2, respectively.

To support uncertainty in the original abstract data types, the operations (and semantics) are extended and the data types *uncertain(gpoint)* and its moving counterpart *mungpoint* that represents static and moving points in networks with uncertainty are introduced. The data types with *mungpoint* and *uncertain(gpoint)* can be seen in Table 8.1. The UNCERTAIN kind with *uncertain* type constructor is added in order to support uncertain data types. As one can note in this type constructor, we also need to use an extension of the base (*BASE*) and spatial (*SPATIAL*) types to support uncertainty.

An uncertain graph point belongs to a network, and the uncertainty is represented only inside the same route. Given a set $N = \{N_1 = (R_1, J_1), \ldots, N_k = (R_k, J_k)\}$ containing all the networks in the database, the *ungpoint* data type is represented as

$$D_{ungpoint} = \{(i, rid, side, pos)\} \bigcup \{\bot\} |$$

$$1 \leq i \leq k,$$

$$pos \in \underline{unreal},$$

$$\exists(rid, len, cc, kind, sm) \in R_i, \text{ such that}$$

$$kind = simple \Leftrightarrow side = none \wedge$$

$$\forall p \in pos, 0 \leq p \leq len$$

There are two approaches for representing the geometry of a moving uncertain graph point as a *mungpoint* data type, i.e., representing a moving object as a set of slices, called *temporal units*. Within each slice, the development of the value can be represented by a temporal function. For example, for the *mreal* data type, a quadratic polynomial function or the secure root of such is used, and for the *mpoint* data type, just a simple linear function.

The temporal functions are represented by the generic function τ that evaluates the unit function at a given time instant. Given a nontemporal type α, its corresponding unit type $D_{u\alpha} = Interval(Instant) \times S_\alpha$, where S_α is a suitably defined set and an $Interval(T)$ is an interval over a set $(U, <)$ with a total order with the following definition:

$$Interval(U) = \{(s, e, lc, rc)\} | s, e \in U, lc, rc \in \underline{bool},$$

$$s \leq e, (s = e) \Rightarrow (lc = rc = T)\}$$

where s and e define the boundaries of the interval and lc and rc define whether the interval is right and/or left-closed. The function τ_α or simply τ is defined as

$$\tau_\alpha = S_\alpha \times Instant \rightarrow D_\alpha$$

Another approach is a slightly modified version of the τ function that receives also as argument the time interval of the unit, i.e., the whole unit. The new τ function is then represented as

$$\tau_\alpha = D_{u\alpha} \times Instant \rightarrow D_\alpha$$

The use of the τ function should become clearer when we instantiate it in the two approaches to represent the moving graph point object.

8.3   Uncertain Trajectory Management

In this section, we first model a road network framework for the UTR-tree hybrid index structure, as well as the network-constrained moving objects and uncertain trajectories. Based on this uncertainty model, we afterwards describe the database operations for uncertainty management.

For simplicity, we model the whole transportation network as a single graph, and we will use "transportation network" and "transportation graph" interchangeably.

8.3.1   Uncertain Trajectory Modeling

Definition 8.1. Motion vectors are snapshots of a moving object's movements and are generated by location updates. A motion vector, mv, is defined as follows:

$$mv = (t, (rid, \, pos), \mathbf{v}) \tag{8.7}$$

where t is a time instant, (rid, pos) is a network position describing the location of the moving object at time t, and $\mathbf{v}$ is the speed measure of the moving object at time t.

The speed measure $\mathbf{v}$ is a real number, which contains both speed and direction information. Its abstract value is equal to the speed of the moving object at time t, while its sign (either positive or negative) indicates the traffic flow direction the moving object belongs to. If the moving object is moving from 0-end towards 1-end, then the sign is positive. Otherwise, if it is moving from 1-end to 0-end, the sign is negative.

In network-constrained moving objects databases, three kinds of location updates are defined in [5], that is, ID-Triggered Location Update (IDTLU), Distance-Threshold Triggered Location Update (DTTLU), and Speed-Threshold Triggered Location Update (STTLU). DTTLU and STTLU are triggered when the moving object exceeds the distance threshold ξ and the speed threshold ψ, respectively, and will generate one motion vector mv_a; IDTLU is triggered when the moving object transfers from one route r_s to another route r_e, and will generate three motion vectors, mv_{a1}, mv_{a2}, mv_{a3}, where mv_{a1} and mv_{a2} correspond to the junction location and mv_{a3} corresponds to the location when IDTLU is triggered. These three kinds of location updates work together to complete the trajectory data sampling process.

Definition 8.2. The trajectory of a moving object mo is a sequence of motion vectors sent by mo through location updates during its journey. A trajectory, denoted as Tr, is defined as follows:

$$Tr = (mv_i)_{i=1}^{n} = ((t_i, (rid_i, pos_i), \mathbf{v}_i))_{i=1}^{n} \tag{8.8}$$

Fig. 8.4 Uncertain trajectory
units of moving objects. (**a**)
Non-active UT-Unit. (**b**)
Active UT-Unit(pentagon).
(**c**) Active UT-Unit(triangle)

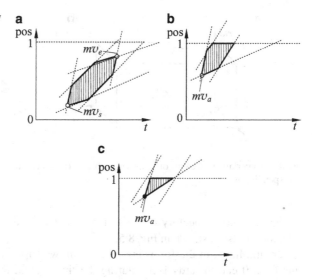

where mv_n is the last motion vector submitted by the moving object, and we call it
the "active motion vector" of the moving object, which contains the key information
for computing the current or near-future locations of the moving object and for
triggering the next location update.

As stated in [6], through the trajectory, we can only know the exact location
of the moving object at the location update time. Between two location updates
and after the last location update, the location of the moving object is uncertain,
and we can only compute the possible locations according to the corresponding
motion vectors. Therefore, the trajectory of the moving object actually describes
the "uncertain locations" of moving objects, and therefore we call it "uncertain
trajectory" in this chapter. For simplicity, trajectory and uncertain trajectory will
be used interchangeably throughout this chapter.

In [2, 6], the authors have analyzed the possible locations that can be derived
from the trajectories. Between any two consecutive motion vectors mv_s and mv_e,
the possible locations of the moving object mo form a hexagon in the $POS \times T$
plane. After the last motion vector mv_a, we can predict the possible location of the
moving object until the end (either 0-end or 1-end, depending on the direction of
mo) of the route, and the possible locations of the moving object form a pentagon,
quadrangle, or triangle, depending on the distance threshold ξ, speed threshold
ψ, and the active motion vector mv_a. We call the above-mentioned polygons
"uncertain trajectory unit," or UT-Unit for short. The UT-Unit corresponding to
two consecutive motion vectors mv_s and mv_e is called non-active UT-Unit and
is denoted as UT-Unit(mv_s, mv_e), and the UT-Unit corresponding to the active
motion vector $mv_a = (t_a, (rid_a, pos_a), \mathbf{v}_a)$ is called active UT-Unit and is denoted as
UT-Unit(mv_a). Figure 8.4 illustrates the geometry of the possible locations derived
from UT-Units.

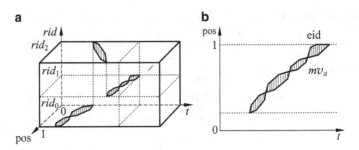

Fig. 8.5 Uncertain trajectory of a moving object. (**a**) Uncertain trajectory *Tr*. (**b**) Part of *Tr* corresponding to a route

The uncertain trajectory of a moving object is a set of consecutive uncertain trajectory units, as shown in Fig. 8.5.

For simplicity, in the following discussion, we suppose that the moving object runs from 0-end towards 1-end during the time period corresponding to the UT-Unit. The methods proposed can be easily adapted to deal with the situation when moving objects run from 1-end to 0-end.

Let us first analyze non-active uncertain trajectory units. From the location update strategies for network-constrained moving objects [5], we know that for a non-active trajectory unit UT-Unit(mv_s, mv_e) (where $mv_s = (t_s, (rid_s, pos_s), \mathbf{v}_s), mv_e = (t_e, (rid_e, pos_e), \mathbf{v}_e)$, and $rid_s = rid_e$), the location of the moving object at any given time $t_q \in [t_s, t_e]$, denoted as $pos[t_q]$, should meet the following condition:

$$pos[t_q] \in [pos_{qmin}, pos_{qmax}] \tag{8.9}$$

where pos_{qmin}, pos_{qmax} can be computed in the following way [6] (since we use relative position $pos \in [0, 1]$, we suppose that the speed measures $\mathbf{v}_s, \mathbf{v}_e, \mathbf{v}_a$, the distance threshold ξ, and the speed threshold ψ have already been divided by *r.length*, the length of the route, before computation):

$$pos_{qmin} = max(pos_q^\diamond - \xi, pos_s + (\mathbf{v}_s - \psi)$$
$$\times (t_q - t_s), pos_e - (\mathbf{v}_s + \psi) \times (t_e - t_q)) \tag{8.10}$$

$$pos_{qmax} = min(pos_q^\diamond - \xi, pos_s + (\mathbf{v}_s + \psi)$$
$$\times (t_q - t_s), pos_e - (\mathbf{v}_s - \psi) \times (t_e - t_q)) \tag{8.11}$$

where $pos_q^\diamond = pos_s + \mathbf{v}_s \times (t_q - t_s)$.

We can depict the geometry of UT-Unit(mv_s, mv_e) as a hexagon (see the shadowed part of Fig. 8.6). As shown in Fig. 8.6, the hexagon of UT-Unit(mv_s, mv_e) is actually formed and surrounded by six lines corresponding to Eqs. (8.10) and (8.11).

Fig. 8.6 Geometry and MBR
of UT-Unit(mv_s, mv_e)

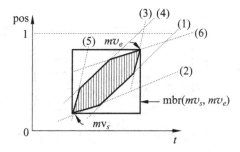

Fig. 8.7 Geometry and MBR
of UT-Unit(mv_a)

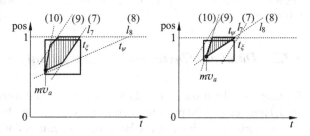

Now let us analyze the MBR of this hexagon. Suppose that $(t*, pos*)$ is an arbitrary point inside the hexagon, where $t_s \leq t* \leq t_e$. Since the moving object runs along $route(rid_s)$ monotonically from pos_s towards pos_e during the time period from t_s to t_e, we can assure that $pos*$ must meet the following condition: $pos_s \leq pos* \leq pos_e$. Therefore, the MBR of UT-Unit(mv_s, mv_e) is: $mbr(mv_s, mv_e)=< t_s, pos_s, t_e, pos_e >$, as shown in Fig. 8.6.

In the following, let us consider active trajectory units. For an active trajectory unit UT-Unit(mv_a), the location of the moving object at a given time $t_q (t_a \leq t_q \leq t_{now})$, denoted as $pos[t_q]$, should meet the following condition:

$$pos[t_q] \in [pos_{qmin}, pos_{qmax}] \tag{8.12}$$

where pos_{qmin} and pos_{qmax} can be computed in the following way [6]:

$$pos_{qmin} = max(pos_q^\diamond - \xi, pos_a + (\mathbf{v}_a - \psi) \times (t_q - t_a)) \tag{8.13}$$

$$pos_{qmax} = min(1, pos_q^\diamond + \xi, pos_a + (\mathbf{v}_a + \psi) \times (t_q - t_a)) \tag{8.14}$$

where $pos_q^\diamond = pos_a + \mathbf{v}_a \times (t_q - t_a)$.

The geometry of UT-Unit(mv_a) is illustrated in Fig. 8.7 (see the shadowed parts).

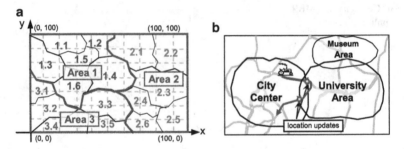

Fig. 8.8 Partition of the underlying traffic network. (**a**) Partition of traffic network. (**b**) Overlap of partition areas

8.3.2   Database Operations for Uncertainty Management

By adjusting the distance threshold ξ and the speed threshold ψ, the uncertainty model presented in the previous section can support variable precisions in presenting the locations of moving objects. However, sometimes this is still "too precise." In a lot of cases, much lower precisions in presenting the locations of moving objects (for instance, "from 8:00 to 10:00, I was traveling in the city center; after that until 13:00 I was visiting the museum area; and from 13:00 to 15:00 I was in the university area") are quite acceptable.

To better support variable precisions in presenting the locations of moving objects, we define a new data type, discretely moving graph region, to describe the possible locations of moving objects.

Definition 8.3. A discretely moving graph region is defined as a sequence of the following form:

$$dmgr = ((t_i, gregion_i))_{i=1}^n$$

where t_i is a time instant and $gregion_i$ is a graph region value. For $\forall i \in \{1, \cdots, n-1\}$, $t_i < t_{i+1}$, and the moving object is assumed to move inside $gregion_i$ between t_i and t_{i+1}.

The locations of moving objects can be tracked in the following way. First, the whole transportation network is partitioned into a group of areas with each area to be a graph region. To support multiple granularity in uncertainty management, the system can have multiple partitions on the same traffic network, which form a hierarchical structure, as shown in Fig. 8.8a. The graph regions are uniquely numbered, and both the server and the moving object need to keep the partition information. To avoid frequent location updates when moving objects are moving near the border between two partitions, these partition areas should overlap each other to some extent, as shown in Fig. 8.8b.

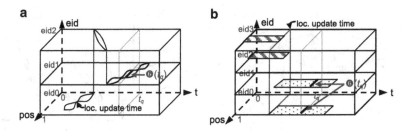

Fig. 8.9 "Trajectories" of moving objects. (**a**) A moving graph route section. (**b**) A discretely moving graph region

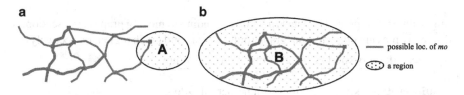

Fig. 8.10 Semantics of (**a**) **inside_possibly** and (**b**) **inside_definitely**

In [9], Güting RH et al. have defined a rich set of operations for network-constrained moving objects and the related data types. Through some extension, these operations can be upgraded to support uncertainty. In this section, we aim to present some general ideas behind the design of the operations and query processing. First let us see the "trajectories" of moving objects. If a moving object is modeled directly in the Euclidean space, its trajectory is a curve (or a tube when uncertainty is considered) in the $X \times Y \times T$ space [10, 16]. However, in network-constrained moving objects databases, the trajectory of a moving object can have totally different forms, as shown in Fig. 8.9.

As illustrated in Fig. 8.9, the possible location of a moving object at any given time instant t_q, denoted by $\omega(t_q)$, is the intersection of its trajectory and the plane corresponding to t_q (which is vertical to the axis). $\omega(t_q)$ can be either a graph route section (see Fig. 8.9a) or a graph region (see Fig. 8.9b).

Let $\partial\omega(t_q)$ be the set of points contained in $\omega(t_q)$. In designing an operation which involves the whole trajectory or part of the trajectory, say, **inside**, we need to extend it to **inside_possibly** and **inside_definitely** with the following semantics:

inside_possibly$(\omega(t_q), A) \Leftrightarrow \exists p \in \partial\omega(t_q)$:**inside**$(p, A) \Leftrightarrow$**intersects**$(\omega(t_q), A)$
inside_definitely$(\omega(t_q), B) \Leftrightarrow \forall p \in \partial\omega(t_q)$:**inside**$(p, B) \Leftrightarrow$ **inside**$(\omega(t_q), B)$
(Fig. 8.10).

Following the strategy described above, we can extend other operations, such as **at**, **intersect**, **atperiod**, and **atinstant**, with uncertainty involved. For instance, the signature of the **atperiod** and **atinstant** operations can be extended as follows

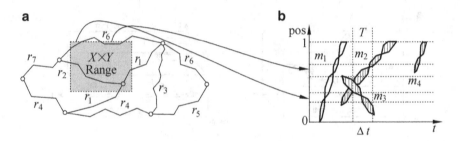

Fig. 8.11 Range query through UTR-tree. (**a**) Search the upper R-tree and receive (*rid* × *period*) pairs. (**b**) Search the upper R-tree and output moving *obj*.identifiers (m_2, m_3)

(mgrs, dmgr, grs, gr are the data types corresponding to moving graph route section, discretely moving graph region, graph route section, and graph region, respectively):

atperiod: mgrs×period→mgrs,dmgr×period→dmgr

atinstant: mgrs×period→mgrs,dmgr×period→dmgr

Since in moving objects databases, the most common uncertain query operators, such as **possibly-inside** (*trajectory*, $Ix \times Iy \times It$) and **possibly-intersect** (*trajectory*, $Ix \times Iy \times It$) (where Ix, Iy, It are intervals in X, Y, T domains), belong to range queries, that is, the input of the query is a range in the $X \times Y \times T$ space, we take range query as an example to show how the uncertain query processing is supported by the UTR-tree [4].

The querying of the UTR-tree can be completed in two steps. When processing a range query (suppose the range is $Ix \times Iy \times It$), the system will first query the upper R-tree of the UTR-tree according to $Ix \times Iy$ and will receive a set of (*rid* × *period*) pairs as the result, where *period* $\subseteq$ [0, 1] and can have multiple elements; then for each (*rid* × *period*) pair, search the corresponding lower R-tree to find the UT-Units intersecting *period* × It, and output the corresponding moving object identifiers. Figure 8.11 illustrates the range query processing based on the UTR-tree. The query algorithm is given in Algorithm 21.

Algorithm 21: RNE (*node_id, QS, result*)

---

input : $Ix \times Iy \times It$
output: Result
Search the upper R-Tree according to $Ix \times Iy$, and receive a set of pairs: $(rid_i, period_i)_{i=1}^{n}$;
for $i:=1$ *to* n **do**
    for $\forall \rho \in period \times It$ **do**
        Let μ be the set of UT-Units in $RTree_{low}(rid_i)$ which intersect ρ;
        $Result = Result \cup$ the set of moving object IDs in the element of μ;
    end
end
Return Result;

8.4   Summary

Uncertainty management is a key research issue with moving objects databases, and a lot of research has been focused on this problem recently. However, most studies focus on the modeling (data types, operations, and algorithms) of uncertainty, leaving the index of uncertain trajectories for moving objects, especially network-constrained ones, as an unsolved problem. In this chapter, we firstly present three representative uncertainty models and then discuss uncertainty management on road networks with two subsections: uncertainty modeling and uncertainty operations. Meanwhile, we present how to process the uncertainty query on this framework. Some other techniques related with uncertainty indexing are also described in other chapters.

References

1. Almeida VT, Güting RH (2005) Indexing the trajectories of moving objects in networks. GeoInformatica 9(1):30–66
2. Almeida VT, Güting RH (2005) Supporting uncertainty in moving objects in network databases. In: Proceedings of the 13th annual ACM international workshop on geographic information systems (GIS 2005), Bremen, pp 31–40
3. Chen J, Meng X (2007) Indexing future trajectories of moving objects in a constrained network. J Comput Sci Technol 22(2):245–251
4. Ding Z (2008) UTR-tree: an index structure for the full uncertain trajectories of network-constrained moving objects. In: 9th international conference on mobile data management (MDM 2008), Beijing, pp 27–30
5. Ding Z, Güting RH (2004) Managing moving objects on dynamic transportation networks. In: Proceedings of the 16th international conference on scientific and statistical database management (SSDBM 2004), Santorini Island, p 287
6. Ding Z, Güting RH (2004) Uncertainty management for network constrained moving objects. In: Proceedings of the 2004 international conference on database and expert systems applications (DEXA 2004), Zaragoza, pp 411–421
7. Frentzos E (2003) Indexing objects moving on fixed networks. In: Proceedings of the 8th international symposium of advances in spatial and temporal databases (SSTD 2003), Santorini Island, pp 289–305
8. Gowrisankar H, Nitte S (2002) Reducing uncertainty in location prediction of moving objects in road networks. In: Proceedings of the 2002 conference on geographic information science (GIScience 2002), Boulder, pp 228–242
9. Güting RH, de Almeida VT, Ding Z (2005) Modeling and querying moving objects in networks. VLDB J 15(2):165–190
10. Pfoser D, Jensen CS (1999) Capturing the uncertainty of moving object representations. In: Proceedings of the 6th international symposium on advances in spatial databases (SSD 1999), Hong Kong, pp 111–132
11. Pfoser D, Jensen CS, Theodoridis Y (2000) Novel approaches in query processing for moving object trajectories. In: Proceedings of the 26th international conference on very large data bases (VLDB 2000), Cairo, pp 395–406
12. Pfoser D, Tryfona N (2001) Capturing fuzziness and uncertainty of spatiotemporal objects. In: Proceedings of the 5th East European conference on advances in databases and information systems (ADBIS 2001), Vilnius, pp 112–126

13. Saltenis S, Jensen CS, Leutenegger ST, Lopez MA (2000) Indexing the position of continuously moving objects. In: Proceedings of the 2000 ACM SIGMOD international conference on management of data (SIGMOD 2000), Dallas, pp 331–342
14. Tøssebro E, Nygård M (2002) Uncertainty in spatio-temporal databases. In: Proceedings of the 2nd international conference on advances in information systems (ADVIS 2002), Izmir, pp 43–53
15. Trajcevski G, Wolfson O, Cao H, Lin H, Zhang F, Rishe N (2002) Managing uncertain trajectories of moving objects with domino. In: Proceedings of the 4th international conference on enterprise information systems (ICEIS 2002), Ciudad Real, pp 769–771
16. Trajcevski G, Wolfson O, Chamberlain S, Zhang F (2002) The geometry of uncertainty in moving objects databases. In: Proceedings of the 8th international conference on extending database technology: advances in database technology (EDBT 2002), Prague, pp 233–250

Chapter 9
Statistical Analysis on Moving Object Trajectories

Abstract Traffic behavior analysis based on moving object trajectories is a basic technique for intelligent transportation system (ITS) applications like traffic control. In this chapter, we firstly propose a new model for objects moving on dynamic transportation networks (MODTN). In the MODTN system, moving objects are modeled as moving graph points that move only within predefined transportation networks. To express general events of the system, such as traffic jams, temporary constructions, and insertion and deletion of junctions or routes, the underlying transportation networks are modeled as dynamic graphs so that the state and the topology of the graph system at any time instant can be tracked and queried. Based on this model, we secondly introduce a real-time traffic flow statistical analysis method called NMOD-TFSA. By analyzing the spatio-temporal trajectories of moving objects, NMOD-TFSA can get the real-time traffic parameter values of the transportation network.

Keywords Statistical analysis • Real-time traffic flow analysis • Dynamic transportation network • Moving object databases

9.1 Introduction

Analyzing or monitoring traffic behavior on transportation network, such as collecting the traffic jam, is an important research direction in the fields of mobile computing and ITS. It is considered very practical to improve traffic conditions, manage transportation systems more effectively, and improve their accessibility. Statistical analysis based on moving object trajectories is recently one of the most used methods to discover the collective behaviors from large scale of individual "moving sensors" (vehicle, pedestrian etc.), hence it has attracted lots of academic researchers and become a key research issue in recent years.

There exist various techniques which have been adopted to collect traffic data, such as stationary sensor-/camera-based methods (monitoring from traffic sensors or

X. Meng et al., *Moving Objects Management: Models, Techniques and Applications*, DOI 10.1007/978-3-642-38276-5_9,
© Tsinghua University Press, Beijing and Springer-Verlag Berlin Heidelberg 2014

optical devices), air-/spaceborne methods (monitoring from airplanes or satellites), and floating-car-based methods (monitoring from floating/probe cars). However, these methods have a lot of limitations. For example, stationary sensor-/camera-based methods can only measure traffic data at fixed positions. To get the traffic information of the whole transportation network, a large number of detectors are needed so that the system can be very expensive. Air-/spaceborne methods can monitor traffic conditions over large areas, but the data are available only when the air-/spaceborne detectors are flying over the monitored areas.

Comparing with above-mentioned traditional methods, a novel kind of method which derives traffic information from the floating-car method [6, 9] (or FCM for short) has attracted increasing research interests in recent years, with a lot of feasible solutions achieved. In the FCM system, certain kinds of vehicles (for instance, taxicabs, buses, or specially equipped probe cars) are equipped with GPS and wireless communication interfaces and periodically (say, once in every 2 min or in every 500 m) report to the central server their locations, velocities, and directions (these data are called floating-car data, or FCD for short). In every certain time interval (say, 5 min), the server launches a statistical process to match the FCD with the traffic network so that traffic flow parameters (for instance, average speed, travel time, and traffic jam of each route) of the network can be computed and refreshed. However, all the above methods are mainly focused on traffic estimation, leaving the balance between data collecting efficiency and statistical analysis accuracy not well studied.

The problems mentioned above are not easy to solve due to two major challenges:

1. How to model the transportation network reasonably? Since the transportation networks can be subject to discrete changes over time, they should be modeled as "dynamic" graphs that allow us to express state changes (such as traffic jams and blockages caused by temporary constructions) and topology changes (such as insertion and deletion of junctions or routes). Correspondingly, moving objects are constrained by the networks in nature; their movement should be also represented as some kind of moving units in the discrete model. In other words, moving objects modeled in the network space are much more complex than them in the European space. Actually, modeling of transportation networks is not trivial as practical situations are really complicated. For example, a logistics network may consist of multiple road graphs, while it may also be organized in different granularity.

2. How to enforce the traffic statistics in real time? First, it needs to determine which kinds of parameters can be used to define the real-time traffic states based on the underlying model. A principle is that these parameters should be suitable both for data sampling and for traffic aware navigation purposes. Second, efficient statistics data structure and access method are needed to speed up the statistical analysis process.

Towards upper considerations, in this chapter, we introduce a novel model of traffic-parameterized road networks and provide several functions and algorithms [3–5] to do statistical analysis of traffic flow in real time.

9.2 Representative Methods

Traditional traffic monitoring techniques based on stationary sensors/cameras require a large number of surface detectors or spaceborne detectors to equip and surely suffer from the challenge of scalability and efficiency. To overcome the above limits, two novel kinds of methods have attracted increasing research interests in recent years, i.e., the methods based on floating-car data (or FCD for short) or based on trajectories in moving object databases (or MOD for short).

9.2.1 Based on FCDs

Currently, a lot of traffic aware solutions are achieved based on FCD data. In order to simplify the discussion, we term the traffic collection methods based on FCDs as FCM for short. FCM is a "moving sensor"-based method as floating cars collect traffic data during their move. In earlier works [16], the authors have analyzed the architecture and the data sampling methods in floating-car systems. In the FCM system, certain kinds of vehicles (for instance, taxicabs, buses, or specially equipped probe cars) are equipped with GPS and wireless communication interfaces and periodically (say, once in every 2 min or in every 500 m) report to the central server their locations, velocities, and directions (i.e., FCD). In every certain time interval (say, 5 min), the server launches a statistical process to match the FCD with the traffic network so that traffic flow parameters (for instance, average speed, travel time, and traffic jam of each route) of the network can be computed and refreshed. In [6, 9], the authors have proposed several data analysis methods for floating-car systems. The paper has analyzed the data sampling frequency for floating cars. In [15], the optimal number of probe cars is analyzed for traffic networks in order to get reasonable statistical results. In [18], the authors have discussed how to derive traffic information through periodically collected GPS data from moving objects. All the above methods are mainly focused on traffic estimation, leaving the balance between data collecting efficiency and statistical analysis accuracy not well studied.

9.2.2 Based on MODs

In recent years, trajectory data collected from large scale of moving objects has greatly enriched the research on data warehouse, OLAP, and data mining. There exist various works on spatio-temporal mining, such as motion pattern discovery [1], trajectory clustering [12], classification [10], outlier detection [11], prediction [14], etc., but most of them are based on Euclidean space. In other words, nearly none of these works have treated the interaction between moving objects and the transportation networks in any way. Euclidean-based solutions are imprecise in describing the network paths the moving objects have taken, because multiple paths over the network can coexist between two consecutive sampling points. These methods are not suitable for mining transportation network-constrained trajectory patterns, so they are hard to be directly used to maintain the traffic network and to optimize the traffic flow control. On the other hand, previous research on moving objects databases is mainly focused on modeling single moving objects [8, 17]. Some recent work has dealt with clustering moving objects to find moving patterns from MOD [7], but the patterns are not as detailed as the real-time traffic parameters as discussed in this section.

More recently, some research has been conducted on traffic flow statistical analysis for trajectories with road network constraints [12, 13]. However, they only consider either the topological features of road networks or the spatio-temporal features of trajectories, and few of them have systematically dealt with more featured traffic patterns, such as uncrowded hot routes, trajectory clusters of resulting traffic blockages, chains of traffic jams, etc.

Recent methods in statistical analysis of traffic flow have a lot of limitations in terms of data sampling costs, data processing efficiency, and statistical analysis accuracy. In other words, existing works would suffer from two shortcomings in practical systems: (1) These limitations make real-time analysis very hard and expensive; (2) the measure error of traffic flow is not trivial because of mismatching between moving object motions and transportation behaviors.

9.3   Real-Time Traffic Analysis on Dynamic Transportation Networks

9.3.1   Modeling Dynamic Transportation Networks

In this section, we first give an example of practical transportation network and then introduce two models of underlying transportation networks. One is the State-Based Dynamic Transportation Network model, which can be used to describe the spatio-temporal aspect of temporally variable transportation networks. The second is the network-constrained moving objects database-based traffic flow statistical analysis (NMOD-TFSA) model, which aims to collect real-time traffic parameter values of the transportation network.

9.3.1.1   An Example of Application Scenarios

Let us first give an example to show how practical transportation networks can be organized in databases. We suppose that the whole system is composed of multiple graphs that can overlap each other. Figure 9.1 gives an example which shows how a modern logistic system works. The whole highway network is expressed as a graph in the database, while the street network of each city is also stored as an independent graph.

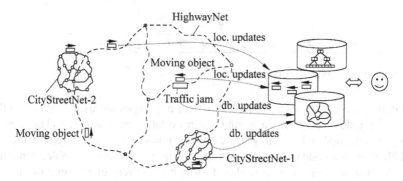

Fig. 9.1 Transportation networks

Each graph can be composed of a set of routes and a set of junctions. The junctions can be further classified into two types: the "in-graph junction" which connects two or more routes of the same graph or "inter-graph junction" which connects multiple graphs. Correspondingly, a certain vehicle can move either by highway between two cities or by street inside a city, during its whole journey. Therefore, it can pass through several different graphs during one trip.

9.3.1.2 The Model of State-Based Dynamic Transportation Network (SBDTN)

The management of moving objects has been intensely investigated in recent years. However, the interaction between moving objects and the underlying transportation networks has been largely ignored. To explore this relationship by involving transportation networks into the modeling of moving objects is one of the main aims of the research project databases for moving objects, which we participate in. In order to meet these requirements, we propose a State-Based Dynamic Transportation Network (SBDTN) model in this section. Its basic idea is to associate a temporal attribute to every node or edge of the graph system so that its state at any time instant can be retrieved. Since the changes of the graph system are discrete, we can use a series of temporal units to represent a temporal attribute with each temporal unit describing one single state during a certain period of time. In this way, the whole state and topology history of the graph system can be presented and queried.

The data model is given as a collection of data types and operations which can be plugged as attribute types into a DBMS to obtain a complete data model and query language. These data types and operations are designed as a discrete model which offers a precise basis for the implementation of data structures in an extensible DBMS such as Secondo [2]. The content of this section mainly focuses on the definitions of graph state, one of the major contributions of Ding and Güting's work [4].

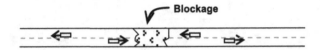

Fig. 9.2 A blocked edge with moving objects

Graph state data types and graph blockage data types are used to describe the
state of a node or an edge. In dynamic transportation networks, a node can have
two states, opened and closed, while an edge can have three states, opened, closed,
and blocked. If a node or an edge is opened, then it is entirely available to moving
objects. If a node or an edge is closed, then it is entirely unavailable to moving
objects, which means that no moving objects are allowed to stay or move in any
part of it. A closed node or edge is not deleted from the system. Instead, it is only
temporarily unavailable to moving objects and can be reopened afterwards.

The blocked state is used to describe a special kind of state of an edge, which
means "partially available" to moving objects. That is, the unblocked part of the
edge is still available to moving objects, but no moving objects can move through
the blocked part. Figure 9.2 gives an example of blocked edge.

Definition 9.1 (State). The carrier set of the *state* data type is defined as follows:

$$D_{state} = \{opened, closed, blocked\}.$$

In a transportation system, blockages can happen quite frequently. For instance,
a road section can be blocked for hours by a car accident or by a temporary
construction, or even by heavy traffic jams. Typically, the location of a blockage
is static. We suppose that the total length of the road section is 1, and then every
location in the road section can be represented by a real number $p \in [0, 1]$. The
location of a blockage can then be expressed as a closed interval over $[0, 1]$, whose
boundaries indicate the border of the blocked area.

Definition 9.2 (Blockage Reason). The data type *blockreason* describes the reason
of a blockage, and its carrier set is defined as follows:

$$D_{blockreason} = \{\text{temporal-construction, traffic-jam, car-accident, others}\}.$$

Definition 9.3 (Interval). Let $(S, <)$ be a set with a total order. Intervals and closed
intervals over S can be defined as follows:

$$interval(S) = \{(s, e, lc, rc)|s, e \in S, lc, rc \in bool, s \le e, (s = e)$$
$$\Rightarrow (lc = rc = true)\}.$$

$$cinterval(S) = \{(s, e, lc, rc)|s, e \in S, lc, s \le e, lc = rc = true\},$$

where lc and rc are two flags indicating "left-closed" and "right-closed," respectively.

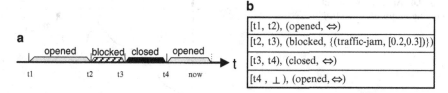

Fig. 9.3 An example temporal attribute value. (**a**) State changes of an edge. (**b**) The corresponding temporal units

Definition 9.4 (Blockage Position).   The data type *blockpos* is used to describe the position of a blockage, and its carrier set is defined as follows:

$$D_{blockpos} = \{\Psi | \Psi \in cinterval([0, 1])\}.$$

In Definition 9.4, we assume that a blockage cannot move during its lifetime. In most cases, this is true. However, sometimes, a blockage can also be dynamic if we take the blockages caused by floods or parades into consideration. In these cases, a blockage should be modeled as a moving interval over [0, 1].

Definition 9.5 (Blockage).   The blockage data type is used to describe a *blockage*, including its reason and its location. The *blockages* data type is used to describe multiple blockages inside one single edge. Their carrier sets are defined as follows:

$$D_{blockage} = \{(br, \Psi) | br \in D_{blockreason}, \Psi \in D_{blockpos}\}$$

$$D_{blockages} = \{B | B \subseteq D_{blockage}\}.$$

Definition 9.6 (State Detail).   The data type *statedetail* is used to describe the detailed state of a node or an edge, and its carrier set is defined as follows:

$$D_{blockages} = \{(s, B) | s \in D_{state}, B \in D_{blockages}, s \neq blocked \Leftrightarrow B = \emptyset\}.$$

Definition 9.6 is based on the fact that several blockages can exist in one road section at the same time so that they should be described as a set of blockages instead of a single blockage value.

We further define the graph temporal data types are used to track the state history and also the life span of a node or an edge in similar way. During its lifetime, a node or an edge can discretely assume a series of states, and each state can last for a certain period of time. In this way, we can decide the topology of the graph system at any time instant. Figure 9.3 illustrates an example temporal attribute value.

In this model, every node or edge of the graph system is associated with a temporal attribute which is composed of a series of temporal units. Each temporal unit describes the state of the node or edge during a certain period of time; hence the model can present the whole state and topology history of the graph system. Much more detail can be found in Ding and Güting's work [4].

Fig. 9.4 Appending current
moving vector mv_{now} to *traj*

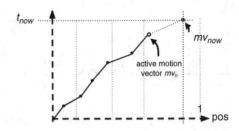

9.3.2 Real-Time Statistical Analysis of Traffic Parameters

As stated earlier, each ARS or junction of the transportation network in NMOD-TFSA has a set of traffic parameters associated to describe its current traffic condition. These basic parameters are refreshed whenever a location update related to the corresponding ARS junction occurs.

Suppose that the functions *route(rid)*, *ars(rid, aid)*, and *junct(jid)* return the route, ARS, and junction corresponding to the identifiers, respectively.

In the following discussion, we first define some trajectory transformation functions, then we provide traffic parameter refreshing algorithms for ARSs and junctions, and finally we describe the statistical data structure and the real-time statistical analysis method in NMOD-TFSA.

9.3.2.1 Trajectory Transformation Functions

Suppose that *traj* is a trajectory, and its last motion vector is $mv_n = (t_n, (rid_n, pos_n), \vec{v}_n, actv_n)$.

The function appcurr(*traj*) appends the current motion vector mv_{now} to the end of *traj*. If the last motion vector of *traj* is active (i.e., $actv_n = true$), then appcurr(*traj*) first computes the location of the moving object at the current time instant t_{now}, denoted as pos_{now}, and then generates a new motion vector $mv_{now} = (t_{now}, (rid_n, pos_{now}), \bot, false)$ and appends it to *traj*, as shown in Fig. 9.4. If the last motion vector of *traj* is inactive (i.e., $actv_n = false$), the function will do nothing.

Function truncate_t(*traj, I*) returns part of *traj* (the result is still a trajectory) which is corresponding to the given time interval $I = [t_1, t_2]$ temporally. Function truncate_g(*traj, ars*) returns part of *traj* which is corresponding to the given atomic route section *ars* geographically. Function truncate_v(*traj*, v_{slow}) returns part of *traj* during which the speed of the moving object is slower than *vslow*. Necessary interpolation may be required to get the end points of the resulted trajectory, as shown in Fig. 9.5.

Function project_t(*traj*) projects *traj* on the time axle and returns a set of time intervals. Function project_g(*traj*) projects *traj* on the geographical plane and returns a set of network route sections.

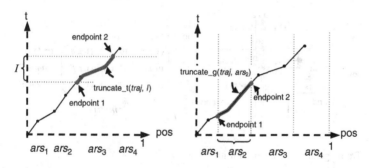

Fig. 9.5 Truncation operators of moving object trajectories

9.3.2.2 Traffic Parameter Refreshing Algorithms for ARSs and Junctions

In this subsection, we consider how to compute traffic parameters through network-constrained moving object trajectories. Let us first consider ARSs. When the traffic parameter refreshing process for a certain directed atomic route section *ars* is triggered, the system will check all trajectories of the moving objects that have stayed in or passed through *ars* in the last Δt time (Δt is a time period of 5–10 min, called statistics window). From each trajectory, the system can derive the travel time and the current position of the corresponding moving object. Therefore, *ars*'s traffic parameters η_{mo} and τ can be computed accordingly.

To get *ars*'s traffic jam status β, the system first needs to compute the jammed area of *ars* as follows:

$$\alpha_{jam} = \bigcap_{i=1}^{n}(project_g(truncate_v(truncate_g(traj_i, ars), v_{slow})))$$

where $\bigcap$ is the spatial intersection operator between network route sections.

From the formula, we can see that α_{jam} is a section of *ars* through which all moving objects move with speed slower than v_{slow} in the last Δt time. Therefore, if α_{jam} is not NULL, then *ars* is blocked. Otherwise, no blockage exists in *ars*. That is:

$$\beta = \begin{cases} true; & (if\ \alpha_{jam} \neq \emptyset) \\ false; & (if\ \alpha_{jam} = \emptyset) \end{cases}$$

The traffic parameter refreshing algorithm for ARSs is given as below:

1. The algorithm first retrieves all the trajectories of moving objects that have passed through *route(rid)* in the last Δt time and gets *TrajSet*.
2. For each trajectory *traj* in *TrajSet*, the algorithm checks whether its latest position is in *route(rid)*. If the position of *traj*'s last moving vector is inside *ars*, adjust η_{mo} accordingly.

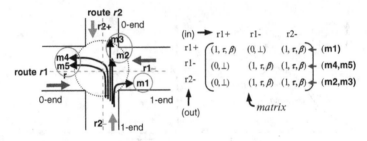

Fig. 9.6 Traffic parameter statistics for junctions

3. Then the algorithm computes the slow speed segments for every trajectory and gets the union of them to α_{jam}.
4. The *jam* status β of *ars* can then be determined from the final result of α_{jam}.
5. We can compute the travel time of each moving object through *ars* by computing two time instants tin and tout (the entering time and exiting time of the moving object on *ars*), so that *ars*'s average travel time τ can be derived.
6. When the statistics is finished, the parameters of *ars* are refreshed with the new values.

The traffic parameter refreshing method for junctions is similar to that of ARSs. For example, η_{mo} can be computed by counting the moving objects whose current position is inside the junction area. The difference is that τ and β need to be computed for each traffic flow inside the junction, as shown in Fig. 9.6.

Suppose that $\xi_{\mu\nu}$ is a traffic flow of the junction *junct* (the in-flow and out-flow are μ and ν, respectively). When computing the average travel time of the junction along $\xi_{\mu\nu}$, denoted as $\tau_{\mu\nu}$, we only need to consider the moving objects running along $\xi_{\mu\nu}$. The traffic jam status of $\xi_{\mu\nu}$, denoted as $\beta_{\mu\nu}$, can be derived from $\tau_{\mu\nu}$. If $\tau_{\mu\nu}$ is longer than a predefined threshold ψ, then $\beta_{\mu\nu} = true$, and otherwise, $\beta_{\mu\nu} = false$.

After that, the in and out traffic flows will be determined, so that the trajectory will only contribute to the statistical computation of the traffic flow it belongs. The parameters are kept in the matrix which is again used to refresh the junction parameters when the statistics is finished for junctions.

9.3.2.3   Statistical Data Structure and Real-Time Traffic Parameter Refreshment

To speed up the statistical analysis, we propose a statistical data structure, called the Current Traffic-status Statistical Analysis Graph (CTSAG), in this subsection. Figure 9.7 illustrates the structure of CTSAG.

As shown in Fig. 9.7, CTSAG includes two B+-Trees, RouteB+-Tree and JunctB+-Tree, which are interconnected with each other through the route and junction records.

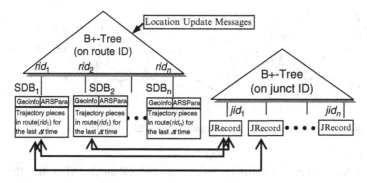

Fig. 9.7 Truncation operators of moving object trajectories

RouteB+-Tree organizes the route records on the *rid* attribute into a B+-Tree structure. The leaf nodes contain records of the form (*rid*, *SDBPointer*), where *rid* is the identifier of the route and *SDBPointer* is a pointer to SDB(*rid*), the statistical data block (SDB) of *route*(*rid*). Each SDB takes the form

$$(geo, len, (aid_i, (jid_{si}, pos_{si}), (jid_{ei}, pos_{ei}), Paraa_i)_{i=1}^n, datasource),$$

where $(geo, len, (aid_i, (jid_{si}, pos_{si}), (jid_{ei}, pos_{ei}), Paraa_i)_{i=1}^n)$ is the route record (with ARS information included) and *datasource* is a set of trajectory pieces acting as the data source for the statistical computation. Each trajectory piece in SDB(*rid*) is still in a trajectory form, but it only contains the motion vectors corresponding to *route*(*rid*). For the sake of efficiency, only the recent Δt time trajectory data corresponding to *route*(*rid*) are kept in *datasource*. Each SDB has a set of pointers $((jpointer_j, pos_j))_{j=1}^m$ leading to the records of junctions within the route.

JunctB+-Tree organizes the junction records on the *jid* attribute into a B+-Tree structure. The leaf nodes contain records of the form (*jid*, *JRecordPointer*), where *jid* is the junction identifier and *JRecordPointer* is a pointer to the junction record of the form (*jid*, *loc*, γ, *matrix*, *Paraj*). Each junction has a set of pointers $((SDBPointer_i, pos_i))_{i=1}^n$ leading to the SDBs of the routes connected by the junction.

When a location update occurs with a moving object *mo*, the system will first save the newly generated motion vector(s) to the corresponding SDB(s) and then refresh the traffic parameters of the related ARSs and junctions by calling traffic parameter refreshing algorithm for ARSs and junctions, respectively, as shown in Fig. 9.8.

Suppose that the last motion vector of *mo* is $mv_n = (t_n, (rid_n, pos_n), \overrightarrow{v}_n, actv_n)$, and the new location update occurs at position (rid_u, pos_u). We notate the geographical path that *mo* has covered from (rid_n, pos_n) to (rid_u, pos_u) as $path_{nu}$.

If *mo* triggers a DTTLU or an STTLU, then a new motion vector $mv_u = (t_u, (rid_u, pos_u), \overrightarrow{v}_u, actv_u)$ will be generated (with $rid_u = rid_n$). In this case, the system will first save mv_u to *mo*'s trajectory piece in SDB(rid_n), and meanwhile,

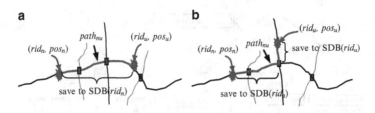

Fig. 9.8 Real-time refreshing at location updates. (**a**) DTTLU or STTLU. (**b**) IDTLU

discard motion vectors in $SDB(rid_n)$ which are older than Δt. After that, the traffic parameters of all ARSs and junctions that intersect $path_{nu}$ will be refreshed, as illustrated in Fig. 9.8a.

If mo transfers from $route(rid_n)$ to $route(rid_u)$ via $junct(jidnu)$ and triggers an IDTLU, then three motion vectors, $mv_{u1} = (t_{u1}, (rid_{u1}, pos_{u1}), \vec{v}_{u1}, actv_{u1})$, $mv_{u2} = (t_{u2}, (rid_{u2}, pos_{u2}), \vec{v}_{u2}, actv_{u2})$, and $mv_{u3} = (t_{u3}, (rid_{u3}, pos_{u3}), \vec{v}_{u3}, actv_{u3})$, will be generated (with $rid_{u1} = rid_n$, $rid_{u2} = rid_{u3} = rid_u$). In this case, the system will save mv_{u1} to mo's trajectory piece in $SDB(rid_n)$ and save mv_{u2} and mv_{u3} to mo's trajectory piece in $SDB(rid_u)$, and then refresh parameters for all ARSs and junctions that intersect $path_{nu}$, as shown in Fig. 9.8b.

Since all the trajectory pieces associated with $route(rid)$ are kept together in $SDB(rid)$, the system can support the trajectories projection and information gathering functions through CTSAG efficiently so that the performance of statistical analysis can be improved.

9.4 Summary

With the recent advancement in mobile computing, sensor networks, and intelligent transportation systems, the network dynamic traffic flow statistical analysis has become a hot research issue. However, current traffic flow analysis methods have a lot of limitations such as high communication costs, low statistical precision, and considerable time delay. To solve these problems, we propose an NMOD-TFSA model in this paper. The experimental results show that compared with floating-car methods which are widely used in real-world applications, NMOD-TFSA provides better performance in terms of data sampling efficiency and statistical precision. In the future work, the traffic aware continuous query based on NMOD-TFSA and the dynamic traffic data broadcasting mechanisms will be dealt with. Also, data warehousing and data mining techniques based on NMOD-TFSA will be studied.

References

1. Benkert M, Gudmundsson J, Hubner F, Wolle T (2008) Reporting flock patterns. Comput Geom Theory Appl 41(3):111–125 (2008)
2. Dieker S, Güting RH (2000) Plug and play with query algebras: SECONDO. A generic DBMS development environment. In: Proceedings of international database engineering and applications symposium (IDEAS 2000), Yokohoma, pp 380–392
3. Ding Z, Güting RH (2004) Managing moving objects on dynamic transportation networks. In: Proceedings of the 16th international conference on scientific and statistical database management (SSDBM 2004), Santorini Island, pp 287–296
4. Ding Z, Güting RH (2004) Modeling temporally variable transportation networks. In: Proceedings of the 9th international conference on database systems for advanced applications (DASFAA 2004), Jeju Island, pp 154–168
5. Ding Z, Huang G (2009) Real-time traffic flow statistical analysis based on network-constrained moving object trajectories. In: Proceedings of the 20th international workshop on database and expert systems applications (DEXA 2009), Linz, pp 173–183
6. Fouladvand M, Darooneh AH (2005) Statistical analysis of floating-car data: an empirical study. Eur Phys J 47(2):319–328
7. Gidofalvi G, Pedersen TB (2009) Mining long, sharable patterns in trajectories of moving objects. Geoinformatica 13(1):27–55
8. Güting RH, De Almeida VT, Ding Z (2006) Modeling and querying moving objects in networks. VLDB J 15(2):165–190
9. Lahrmann H (2007) Floating car data for traffic monitoring. In: Proceedings of the i2TERN conference, Aalborg, June 2007
10. Lee JG, Han J, Li X, Gonzalez H (2008) TraClass: trajectory classification using hierarchical region-based and trajectory-based clustering. Proc VLDB Endow 1(1):441–459. (PVLDB 2008)
11. Lee J, Han J, Li X (2008) Trajectory outlier detection: a partition-and-detect framework. In: Proceedings of the 24th international conference on data engineering (ICDE 2008), Cancun, pp 140–149
12. Li X, Han J, Lee JG, Gonzalez H (2007) Traffic density-based discovery of hot routes in road networks. In: Proceedings of international conference on advances in spatial and temporal databases (SSTD 2007), Boston, pp 441–459
13. Lo CH, Peng WC, Chen CW et al (2008) CarWeb: a traffic data collection platform. In: Proceedings of the 9th international conference on mobile data management (MDM 2008), Beijing, pp 221–222
14. Monreale A, Pinelli F, Trasarti R, Giannotti F (2009) Wherenext: a location predictor on trajectory pattern mining. In: Proceedings of the 15th ACM SIGKDD international conference on knowledge discovery & data mining (KDD 2009), Paris, pp 637–646
15. Park CG, Oh J, Kim S (1998) Determination of optimal number of probe vehicles for real-time traffic flow information. In: Proceedings of the 5th world congress on intelligent transport systems (ITS 1998), Seoul, p 4088
16. Sarvi M, Horiguchi R, Kuwahara M, Shimizu Y (2003) A methodology to identify traffic condition using intelligent probe vehicles. In: Proceedings of the 10th world congress on intelligent transport systems (ITS 2003), Madrid, pp 17–21
17. Speicys L, Jensen CS, Kligys A (2003) Computational data modeling for network-constrained moving objects. In: Proceedings of the 11th ACM international symposium on advances in geographic information systems (GIS 2003), Louisiana, pp 118–125
18. Yoon J, Noble B, Liu M (2007) Surface street traffic estimation. In: Proceedings of the 5th international conference on mobile systems, applications and services, San Juan, pp 220–232

Chapter 10
Clustering Analysis of Moving Objects

Abstract In many moving objects management applications, real-time data analysis such as clustering analysis is becoming one of the most important requirements. Most spatial clustering algorithms deal with objects in Euclidean space. In many real-life applications, however, the accessibility of spatial objects is constrained by spatial networks (e.g., road networks). It is therefore more realistic to work on clustering objects in a road network. The distance metric in such a setting is redefined by the network distance, which has to be computed by the expensive shortest path distance over the network. The existing methods are not applicable to such cases. Therefore, we use the information of nodes and edges in the network to present two new static clustering algorithms that prune the search space and avoid unnecessary distance computations. In addition, we present the problem of clustering moving objects in spatial networks and propose a unified framework to address it. The goals are to optimize the cost of clustering moving objects and support multiple types of clusters in a single application. Furthermore, we introduce two trajectory clustering algorithms in detail: One is that partitions a trajectory into set of line segments and groups similar line segments together into a cluster, and the other is that clusters trajectories based on features other than density; the primary advantage of this clustering is to avoid the big region problem suffered from density-based clustering.

10.1 Introduction

Clustering is one of the most important analysis techniques. It groups similar data to provide a summary of data distribution patterns in a dataset. Early research mainly focused on clustering a static dataset [6–8, 12, 13, 18, 19, 21, 26]. In recent years, there has been increasing research on clustering moving objects [5,11,16,25], which has various applications in the domains of weather forecast, traffic jam prediction,

X. Meng et al., *Moving Objects Management: Models, Techniques and Applications*, DOI 10.1007/978-3-642-38276-5__10,
© Tsinghua University Press, Beijing and Springer-Verlag Berlin Heidelberg 2014

and animal migration analysis, to name but a few. However, most existing work on clustering moving objects assumed a free movement space and defined the similarity between objects by their Euclidean distance.

In the real world, objects move within spatially constrained networks, e.g., vehicles move on road networks and trains on railway networks. Thus, it is more practical to define the similarity between objects by their network distance – the shortest path distance over the network. Therefore, by exploiting unique features of road networks, two new clustering algorithms for static objects are presented, which use the information of nodes and edges in the network to prune the search space and avoid some unnecessary distance computations. For clustering moving objects in road networks, we propose a unified framework for "clustering moving objects in spatial networks" (CMON). Due to the innate feature of continuously changing positions of moving objects, the clustering results dynamically change. By exploiting the unique features of road networks, the CMON framework first introduces a notion of cluster block (CB) as the underlying clustering unit. We then divide the clustering process into the continuous maintenance of CBs and periodical construction of clusters with different criteria based on CBs. The algorithms for efficiently maintaining and organizing the CBs to construct clusters are proposed.

The trajectory of moving object is movements of moving object generated by location updates; an efficient clustering algorithm for trajectories is essential for analysis tasks; therefore, two trajectory clustering algorithms are presented, which use the partition-and-group framework for clustering trajectories and a filter-refinement framework for hot region discovery, respectively.

10.2   Underlying Clustering Analysis Methods

The goal of clustering analysis is to divide a collection of objects into groups, such that the similarity between objects in the same group is high and objects from different groups are dissimilar. In spatial databases, objects are characterized by their position in the Euclidean space, and, naturally, dissimilarity between two objects is defined by their Euclidean distance. Several clustering techniques have been proposed for static datasets in a Euclidean space. They can be classified into the partitioning [13, 21], hierarchical [6, 19, 26], density-based [18], grid-based [1, 22], and model-based [4] clustering methods.

The generic definition of clustering is usually refined depending on the type of data to be clustered and the clustering objective. In other words, different clustering paradigms use different definitions and evaluation criteria. Partitioning methods divide the objects into k groups and iteratively exchange objects between them until the quality of the clusters does not further improve. k-means and k-medoids are representative methods from this class. In k-means algorithms, clusters are represented by a mean value (e.g., a Euclidean centroid of the points in it), and object exchanging stops if the average distance from objects to their cluster's mean value converges to a minimum value. k-medoids algorithms represent each cluster by an actual object in it. First, k-medoids are chosen randomly from the dataset. An evaluation function sums the distance from all points to their nearest medoid. Then, a medoid is replaced by a random object, and the change is committed only if it results in a smaller value of the evaluation function. A local optimum is reached, after a large sequence of unsuccessful replacements. This process is repeated for a number of initial random medoid sets, and the clusters are finalized according to the best local optimum found.

Another class of (agglomerative) hierarchical clustering techniques define the clusters in a bottom-up fashion, by first assuming that all objects are individual clusters and gradually the closest pair of clusters are merged until a desired number of clusters remain. Several definitions for the distance between clusters exist; the single-link approach considers the minimum distance between objects from the two clusters. Others consider the maximum such distance (complete-link) or the distance between cluster representatives. Divisive hierarchical methods operate in a top-down fashion by iteratively splitting an initial global cluster that contains all objects. The cost of brute-force hierarchical methods is $O(N^2)$, where N is the number of objects, which is not suitable for practical use. Moreover, they are sensitive to outliers (like partitioning methods). Algorithms like BIRCH [26] and CURE [6] were proposed to improve the scalability of agglomerative clustering and the quality of the discovered partitions. C2P [19] is another hierarchical algorithm similar to CURE, which employs closest pairs algorithms and uses a spatial index to improve scalability.

Density-based methods discover dense regions in space, where objects are close to each other and separate them from regions of low density. DBSCAN [18] is the most representative method in this class. First, DBSCAN selects a point p from the dataset. A range query, with center p and radius ε, is applied to verify if the neighborhood of p contains at least a minimum number of points, that is, *MinPts* (i.e., it is dense). If so, these points are put in the same cluster as p, and this process is iteratively applied again for the new points of the cluster. DBSCAN continues until the cluster cannot be further expanded; the whole dense region in which p falls is discovered. The process is repeated for unvisited points until all clusters and outlier points have been discovered. A limitation of this approach (addressed in [2]) is that it is hard to find appropriate values for ε and *MinPts*.

In many real applications, however, the accessibility of spatial objects is constrained by spatial (e.g., road) networks. It is therefore realistic to define the dissimilarity between objects by their network distance, instead of the Euclidean distance. The network distance between two objects p and q is defined by the length of the shortest path that reaches q from p and vice versa, assuming an undirected network graph. There are also a few studies [8, 12, 23, 24] on clustering nodes or objects in a spatial network. Jain and Dubes [8] used the agglomerative hierarchical approach to cluster nodes of a graph. They treat each node as a cluster and then merge the clusters until one remains. The single-link variant of this method has complexity $O(|V|^2)$, whereas the complete-link variant comes with complexity $O(|V|^2 \log |V|)$. Both methods are not scalable for large networks.

Another variant [24] applies divisive clustering on the minimum spanning tree of the graph, which can be computed in $O(|V| \log |V|)$ time. However, this method is very sensitive to outliers. CHAMELEON [12] is a general-purpose algorithm, which transforms the problem space into a weighted kNN graph, where each object is connected with its k nearest neighbors. The weight of each edge reflects the similarity between the objects. Yiu and Mamoulis [23] defined the problem of clustering objects based on the network distance. They proposed algorithms for three different clustering paradigms, i.e., k-medoids for K-partitioning, ε-link for density-based, and *single-link* for hierarchical clustering. These algorithms avoid computing distances between every pair of network nodes by exploiting the properties of the network. However, all these solutions assumed a static dataset. A straightforward extension of these algorithms to moving objects by periodical reevaluation is inefficient. Besides, Jin et al. [10] studied the problem of mining distance-based outliers in spatial networks and found the problem to be only a by-product of clustering.

Clustering analysis on moving objects has recently drawn increasing attention. Li et al. [16] first addressed this problem by proposing a concept of micro moving cluster (MMC), which denotes a group of similar objects both at current time and at near-future time. Each MMC is tightly bounded by a rectangle, whose size grows with time. In order to obtain high-quality clusters, the MMCs are kept geographically small. Specifically, they identify the split and merge events and dynamically maintain the bounding boxes of clusters by their width or height of the bounding box. Each MMC maintains a bounding box for the moving objects contained, whose size grows over time. Zhang and Lin [25] proposed a histogram construction technique based on a clustering paradigm. In [11], Kalnis proposed three algorithms to discover moving clusters from historical trajectories of objects. Nehme and Rundensteiner [20] applied the idea of clustering moving objects to optimize the continuous spatio-temporal query execution. The moving cluster is represented by a circle in their algorithms. However, most of the studies only considered moving objects in unconstrained environments and defined the similarity between objects by their Euclidean distance. This chapter specifies the problem of clustering network-constrained moving objects whose similarity is defined by network distance.

10.3   Clustering Static Objects in Spatial Networks

For clustering objects in a spatial network, the distance metric is redefined by the network distance, which has to be computed by the expensive shortest path distance over the network. We presented two new clustering algorithms that use the information of nodes and edges in the network to prune the search space and avoid some unnecessary distance computations.

10.3.1 Problem Definition

In this section, we formally define the problem space to which we apply clustering and the distance metric used in our settings. We introduce the definition of network, network distance between objects, and network distance between clusters. We then identify the peculiarities of the problem by definition of cluster block and discuss why existing clustering algorithms are inapplicable or inefficient for objects that lie on a network.

Definition 10.1. A network is an undirected weighted graph $G = (V, E, W)$ where V is the set of vertices (i.e., nodes), E is the set of edges, and $W : E \rightarrow IR^+$ associates each edge to a positive real number. An object (i.e., point) is located on an edge $e \in E$ in the network. The position of the object in the network can be expressed by the triplet $<n_i, n_j, pos_i>$, where $pos \in [0, W(e)]$ is the distance of the point from node n_i along the edge.

A point lies on exactly one edge. (In real-life problems, some objects may not lie on edges of the network. In such cases, we assume that the object is represented by the position on the network which is most directly accessible from it.) To ensure the position of the object is expressed unambiguously by one triplet, we require that $n_i < n_j$ (assuming a total ordering of node labels).

Let p and q be two points, whose positions are $<n_a, n_b, pos_{p_i}>$ and $<n_c, n_d, pos_{q_i}>$, respectively. The network distance $Dd(n_i, n_j)$ is defined in Definition 10.2.

Definition 10.2. The network distance is defined as follows:

1. The direct distance between points in the same network edge: $Dd(p, q) = | pos_{p_i} - pos_{q_j} |$ (p and q lie on the same edge, e.g., $n_a = n_c$ and $n_b = n_d$); otherwise, it is defined as ∞.
2. The direct distance between a point and a node in the same network edge: $Dd(p, n_a) = pos_p$; $Dd(p, n_b) = W(n_a, n_b) - pos_p$.
3. The network distance between nodes: $Dn(n_i, n_j)$ is the distance of the shortest path from n_i to n_j.
4. The network distance between objects in different network edges: $Dn(p, q) = min_{x \in \{a,b\}, y \in \{c,d\}} Dd(p, n_x) + Dn(n_x, n_y) + Dd(n_y, q)$.

From the definition, in the computation of the network distance of objects, the first two cases are defined as the direct distance, which is not necessarily the network distance and can be found in constant time. However, the last two cases are defined as the network distance, which needs the computation of the shortest distance between nodes.

Definition 10.3. The network distance between clusters is the minimum network distance among different boundary objects of the clusters. Let the boundary objects of the cluster C_x be $p_{x_1}, p_{x_2}, \cdots, p_{x_m}$ and the boundary objects of the cluster C_y be $p_{y_1}, p_{y_2}, \cdots, p_{y_n}$; the similarity of the clusters C_x and C_y, $M(C_x, C_y) = min_{i \subset \{1,m\}, j \subset \{1,n\}} Dn(P_{x_i}, P_{y_j})$.

It is necessary to find the boundary objects of the clusters in the network distance computation of clusters. In the road network, we treat the closest object in the cluster to the nodes as the boundary object of the cluster since the boundary of the clusters is relevant to nodes.

The object cluster results in the road network are usually composed of object sets in a few of network edges. Therefore, for representing the cluster results better and reducing the computation of the network distance of objects, we introduce the definition of cluster block (CB).

Definition 10.4. A cluster block is represented by $(O, n_a, n_b, head, tail, ObjNum)$, where O is a list of objects $\{o_1, o_2, \cdots, o_i, \cdots, o_n\}, o_i = (oid_i, n_a, n_b, pos_i)$. Without loss of generality, assuming $pos_1 \leq pos_2 \leq \cdots \leq pos_n$, it must satisfy $|pos_{i+1} - pos_i| \leq \varepsilon \ (1 \leq i \leq n - 1)$. Since all objects are on the same edge (n_a, n_b), the position of the cluster is determined by an interval *(head, tail)* in terms of the network distance from n_a. Thus, the length of the CB is $|tail\text{-}head|$. *ObjNum* is the number of objects in the CB.

There could be several CBs in the same edge, but each CB only belongs to one edge. A CB itself is a cluster. Therefore, clusters in the network are composed of one or several adjacent CBs. Given a collection of N object points that lie on a network, our objective to group them into a set of clusters becomes constructing the CBs according to their direct distance and merging these CBs by their network distance.

Existing spatial clustering methods group objects only by their spatial similarity, which is either infeasible or inefficient for clustering network-based objects. The replacement of the Euclidean distance by the network distance increases the complexity, since now the distance between two arbitrary objects cannot be computed in constant time, but an expensive shortest path algorithm is required. However, we find that the cluster results in a network are relevant to the edges and nodes of the network. For example, the objects in the same edge or adjacent edges are likely to be grouped into a cluster. Similarly, the chance that the objects around one node are clustered together is large (e.g., traffic jams usually occur in the crossroad). Based on these observations, we exploit the information of edges and nodes and propose two clustering methods, edge-based clustering and node-based clustering, in the following sections.

10.3.2   Edge-Based Clustering Algorithm

Most hierarchical clustering methods initially assume that each point is a cluster and then iteratively merge the closest pair of clusters until one cluster remains. The user may opt to stop the algorithm after a desired number of k clusters have been discovered. Since the algorithm initializes one cluster for each point in the dataset, considerable merging of clusters is involved when the number of objects becomes high. The edge-based clustering algorithm is actually a hierarchical

Fig. 10.1 Initiation phase

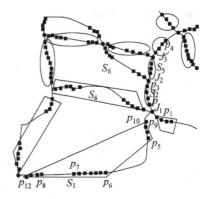

method. However, it solves the scalability issues of the traditional hierarchical clustering methods by constructing initial groups according to the edges in which the objects lie and refining the results through group splitting and merging. During the merging process, the algorithm only merges the clusters adjacent to nodes and further reduces the number of merges by introducing the ε parameter so that the fine-granular cluster results can be found at the earliest.

The edge-based clustering algorithm involves the following three phases:

- Initiation phase: Construct initial groups according to edges in which the objects lie. This involves assigning the objects in the same edge into one cluster. The number of initial clusters is the number of edges in the network. This phase can filter out those edges in which no objects lie. Therefore, unnecessary network traverse processes are reduced.
- Splitting phase: Split large initial groups into smaller cluster blocks to obtain the intermediate results. Specifically, for each initial group, if the network distance of two adjacent objects in this group is larger than the predefined threshold ε, the group needs to be split into two smaller cluster blocks from the two adjacent objects. Otherwise, the group is treated as an individual cluster block. The process ensures the compactness of cluster blocks.
- Merging phase: Iteratively merge the adjacent cluster (blocks) to form the final cluster results. Specifically, for each node, the process merges the adjacent cluster (blocks) around the node iteratively until they cannot merge anymore according to the threshold ε and the network distance of clusters.

Figure 10.1 shows an example of the edge-based clustering process. A part of the road network is represented as road segments (denoted as S) and intersections (denoted as J) in the figure, which correspond to edges and nodes, respectively, in the network graph. Objects (denoted as P) are represented by small rectangles and clusters by circles or polygons. In the initial phase, all objects in the same road segment are clustered into one group. For example, for segment S_1, the objects p_9, p_5, p_6, p_7, p_8, p_{12} and other objects between them are grouped together. Similarly, the cluster in segment S_2 contains all objects between p_2 and p_3. In the splitting phase, since the network distances between p_5 and p_6, as well as between p_7 and

Fig. 10.2 Splitting phase

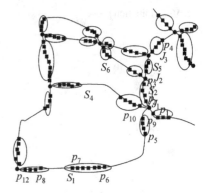

Fig. 10.3 Merging phase

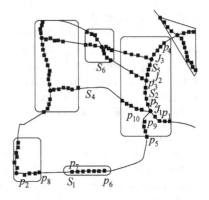

p_8, are larger than the threshold ε, the corresponding cluster is split into three parts as shown in Fig. 10.2. In the merging phase, for cluster blocks around nodes J_1, J_2, and J_3, since the network distances between adjacent cluster blocks are less than the threshold ε, they are merged into a large cluster. Figure 10.3 shows the final results after repeating this merging process.

The pseudo-codes of the edge-based algorithm are shown in Algorithm 22.

To reduce the number of object access in the network, in the initial phase, the algorithm creates clusters for each edge having objects instead of assigning objects to clusters. In the splitting phase, when traversing the objects in the corresponding edge, objects are assigned the split CBs. The key part of the algorithm is merging of the clusters (blocks). After splitting, there are two cases for which clusters need to be merged: (1) clusters adjacent to the nodes and (2) clusters across a small edge having no other clusters but with network distance between them less than the threshold. The algorithm maintains a ε-*nodelist* for each cluster in which each entry is the pair of (ε-*node*, ε-*dist*). ε-*node* is the adjacent node of the cluster satisfying the condition that the network distance between the node and the cluster is less than ε and ε-*dist* denotes the network distance. The algorithm uses the list of ε-*node* for clusters and sorts the clusters by the network distance during merging. This will filter out some unnecessary clusters whose distance to the nodes is larger than ε and reduce the

---

Algorithm 22: Edge_CMON( )

// Q: priority queue which is used to store clusters to be merged around nodes;
// Initial Phase;
for *each network edge (n_x, n_y) in edge-list with moving objects on it* **do**
   | Create a new cluster C for edge (n_x, n_y), assign c_{id} for it;
end
// Splitting Phase;
for *each cluster C_i* **do**
   o is the first object on edge (n_x, n_y) in which C_i lies;
   $o.c_i d = C_i.cid$;
   if $Dd(o.pos, nx) \leq \varepsilon$ **then**
     | Insert $< nx, Dd(o.pos, nx) >$ into nodelist of C_i;
   end
   nexto is the next object on edge (n_x, n_y) from o to n_y; $C = C_i$;
   while *nexto is not NULL* **do**
     if $Dd(o.pos, nexto.pos) > \varepsilon$ **then**
       Split C into CB_1 and CB_2; $C = CB_2$;
       if *nexto is the last object on edge (n_x, n_y) AND $Dd(nexto.pos, n_y) \leq \varepsilon$* **then**
         | Insert $< ny, Dd(nexto.pos, ny) >$ into nodelist of C_i;
       end
       $o = nexto$;
       nexto is the next object on edge (n_x, n_y) from o to n_y;
       $o.cid = C.cid$;
   end
   // Merging Phase;
   for *each n_i in nodelist of C_i* **do**
     Q =new priority queue;
     Insert clusters in n_i into Q according to their distance to n_i;
     if *notempty(Q)* **then**
       C_i is the first cluster in Q; C_j is the next cluster in Q;
       while $(Dd(C_i, n_i) + Dd(C_j, n_i)) \leq \varepsilon$ **do**
         Merge C_j into C_i and merge the nodelist of C_j into the one of C_i;
         if *notempty(Q)* **then**
           | C_j is the next cluster in Q;
         end
         else
           | Break;
       end
     for *each adjacent node n_s of n_i* **do**
       if *no cluster in edge (n_i, n_s) and $(Dd(Ci, ni) + W(ni, ns)) \leq \varepsilon$* **then**
         | Insert $< ns, (Dd(C_i, n_i) + W(n_i, n_s)) >$ into nodelist of C_i;
       end
     end
   end
end
end

---

repeated computation of the distance change of clusters due to the distance change between clusters and nodes after merging.

10.3.3   Node-Based Clustering Algorithm

Given a random point p, the density-based clustering method identifies the cluster, to which p belongs, by applying an ε-range query around p and checking if there are at least *MinPts* points in this range. If so, a new cluster for p is created containing the points in the range query. It iteratively applies range queries for the new points in the cluster, until it cannot be expanded any further. The node-based clustering algorithm involves adapting this density method to our network model. A main module of the algorithm finds the ε-neighborhood of a point p in the network. This can be done by expanding the network around p and assigning points until the distance exceeds ε.

The node-based clustering algorithm further optimizes the network-expanding process by exploiting the nodes information of the network, which avoids the redundant computation of random expanding. The main idea is to traverse the network starting from the node and the group objects around the node according to the condition that their network distance is less than ε. Then, the algorithm expands the cluster to the adjacent nodes so that the other objects around the adjacent nodes are also grouped into this cluster when these objects satisfy the same condition as well. The process continues until the cluster cannot expand (e.g., the distance between any adjacent object and the cluster exceeds ε). For other nodes which are not traversed, we repeat this process until all objects around the nodes are assigned to some cluster. Finally, we check the isolated objects that cannot join some cluster to check whether they can form the individual clusters.

The core part in the node-based clustering algorithm is to cluster objects around one node and expand to other adjacent nodes. The clustering process based on one node can be divided into two steps: initial phase and expanding phase. In the initial phase, the algorithm first filters out the edges containing the node in which the distance between objects and the node is larger than ε. Then, the objects satisfying the distance condition are sorted by the distance to the node. In this way, the nearest object to the node is treated as an initial cluster. During the expanding phase, we expand the initial cluster according to the ordered objects in the adjacent edges. Then we continue the sorting and expanding process for the adjacent nodes until the network distance between adjacent objects is larger than ε and the cluster cannot expand again. The algorithm is shown in Algorithm 23.

Figure 10.4 shows an example of the node-based clustering process, which starts from the node J_1. The objects around J_1 are traversed and ordered according to their distance to J_1. Then, the initial cluster expands from J_1 to the adjacent node J_2 until the next adjacent node J_3. Consequently, the objects around these nodes are

Algorithm 23: Node_CMON( )

// Q: priority queue in which each entry B is $< node1, node2, dist >$. This represents that the distance between $node1$ and adjacent objects in edge ($node1$,$node2$) is $dist$;
// $Ndist[ni]$: store the distance of the nearest object in adjacent edges to ni. Default value is set to ∞;

for *each adjacent node nz to ni* **do**
    o is the first object in edge (ni,nz);
    if *(o is not Null AND $Dd(o.pos, ni) \leq \varepsilon$) OR (o is NULL AND $W(ni, nz) \leq \varepsilon$)* **then**
        Insert $< ni, nz, Dd(o.pos, ni) >$ or $< ni, nz, W(ni, nz) >$ into Q by the distance between o or nz to ni;
    end
end
if *Notempty(Q)* **then**
    Create a new cluster C, assign cid for it;
end
// Expanding Phase;
while *Notempty(Q)* **do**
    B = Dequeue(Q); $nx=B.node1$; $ny = B.node2$;
    if *$B.dist < Ndist[nx]$* **then**
        $Ndist[nx] = B.dist$;
    end
    o is the first object in edge (nx, ny);
    if *$Dd(o.pos, nx) + Ndist[nx] \leq \varepsilon$ OR o is the closest object to nx* **then**
        $o.cid = C.cid$; clustered(o)=True;
        nexto is the next object in edge (nx, ny) from o to ny;
        while *nexto is not Null AND $Dd(o.pos, nexto.pos) \leq \varepsilon$* **do**
            $nexto.cid=C.cid$; Clustered($nexto$)=True;
            $o = nexto$; $nexto$ is the next object in edge (nx, ny) from o to ny;
            if *$Dd(o.pos, ny) \leq \varepsilon$* **then**
                if *$Dd(o.pos, ny) < Ndist[ny]$* **then**
                    $Ndist[ny] = Dd(o.pos, ny)$;
                end
            visited(ny)=True;
            for *each adjacent node nz (except nx) to ny* **do**
                o is the first object in edge(ny, nz);
                if *o is not Null AND $Dd(o.pos, ny) \leq \varepsilon$ OR (o is Null AND $Ndist[ny]+W(ny, nz) \leq \varepsilon$)* **then**
                    Insert $< ny, nz, Dd(o.pos, ny) >$ or $< ny, nz, Ndist[ny] + W(ny, nz) >$ into Q by the distance between o or nz to ny;
                end
            end
        end
    end
end

Fig. 10.4 Cluster of the
node J_1

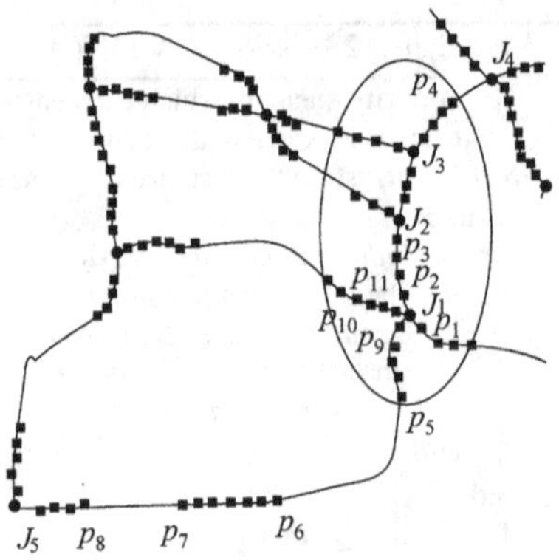

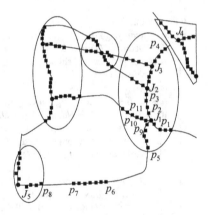

Fig. 10.5 Clusters of all
nodes

traversed and assigned to this cluster if they satisfy the distance condition. When
the cluster cannot expand anymore, the algorithm selects other nodes that are not
traversed and repeats this process until all nodes are traversed (shown as Fig. 10.5).
Finally, it checks the individual objects such as p_6, p_7, and objects between them to
see whether they can form an individual cluster according to the distance between
each pair of adjacent objects. The final clustering results in this example are shown
in Fig. 10.6. If the user sets the minimum object number in a cluster to be five, for
example, the objects p_6, p_7, and the ones between them are treated as outliers.

The algorithm needs a priority queue Q to keep the adjacent edges of a node and
network distance of objects adjacent the node. In Q, edges are grouped by nodes and
sorted according to their distance to each node. The traversed nodes in the network-
expanding process are inserted into the head of Q. The array *Ndist* keeps the nearest
distance of adjacent objects to each node to decide which adjacent edge needs to be
traversed and which objects need to be added into initial clusters. The algorithm
also deals with the case of clusters across a small edge having no other clusters but
with network distance between them less than the threshold. In this case, the objects
need to be added into initial clusters and expanded.

Fig. 10.6 Clusters of all
objects

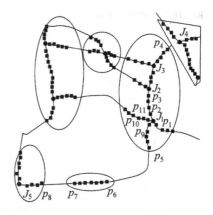

10.4   Clustering Moving Objects in Spatial Networks

Clustering moving objects in spatial networks is more complex than in free space.
The increasing complexity is mainly due to the network distance metric. The
distance between two arbitrary objects cannot be obtained in constant time, but
requires an expensive shortest path computation. Moreover, the clustering results
are related to the segments of the network, and their changes will be affected by the
network constraint. For example, a cluster is likely to move along the road segments
and change (i.e., splitting and merging) at the road junctions due to the objects'
diversified spatio-temporal properties (e.g., moving in different directions). It is not
efficient to predict their changes only by measuring their compactness. Thus, the
existing clustering methods for free space cannot be applied to spatial networks
efficiently.

On the other hand, the existing clustering algorithms based on the network
distance [23] mainly focus on the static objects that lie on spatial networks. To
extend to moving objects, we can apply them to the current positions of the objects
in the network periodically. However, this approach is highly expensive since
each time the expensive clustering evaluation starts from scratch. In addition, the
clustering algorithms for different clustering criteria (e.g., K-partitioning, distance,
and density based) are totally different in their implementation. This is inefficient
for many applications that require to execute multiple clustering algorithms at the
same time. For example, in a traffic management application, it is important to
monitor densely populated areas (by density-based clusters) so that traffic control
can be applied, but at the same time, there may be a requirement for assigning K
police officers to each of the congested areas. In this case, it is favorable to partition
the objects into K clusters and keep track of the K-partitioned clusters. Separate
evaluation of different types of clusters may incur computational redundancy.

In this section, we introduce a unified framework for clustering moving objects in
spatial networks (CMON). The goals are to optimize the cost of clustering moving
objects and support multiple types of clusters in a single application. The CMON

Fig. 10.7 CMON framework

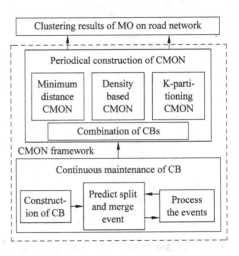

framework divides the clustering process into the continuous maintenance of cluster blocks (CBs) and the periodical construction of clusters with different criteria based on CBs. A CB groups a set of objects on a road segment in close proximity to each other at present and in the near future. In general, a CB satisfies two basic requirements: (1) It is inexpensive to maintain in a spatial network setting; (2) it serves as a building block for different types of application-level clusters.

10.4.1 CMON Framework

We model a spatial network as a graph where objects are moving on the edges (we use the word "segments" for "edges" interchangeably). The distance between any two objects, called *network distance*, is measured by the length of the shortest path connecting them in the network. We employ a similar motion model as in [16], where moving objects are assumed to move in a piecewise linear manner (i.e., each object moves at a stable velocity at each edge). We assume that an object location update has the following form $(oid, n_a, n_b, pos, speed, next\_node)$, where *oid* is the *id* of the moving object, (n_a, n_b) represents the edge on which the object moves (from n_a towards n_b), *pos* is the relative location to n_a, and *speed* is the moving speed. We also assume that the next edge to move along, $(n_b, next\_node)$, is known in advance. The requirement is to continuously monitor the moving clusters with various criteria at some predefined period.

As shown in Fig. 10.7, the proposed CMON framework is composed of two components: the incremental maintenance of cluster blocks (CBs) and the periodical construction of different types of application-level clusters. We have defined a CB in Definition 10.4. A CB is a group of moving objects close to each other at present and near-future time. For easy maintenance, we constrain the objects in a

CB moving in the same direction and on the same edge segment. Additionally, a CB imposes a strict clustering criterion so as to support different types of application-level clusters. Specifically, the network distance between the pairs of neighboring objects in a CB does not exceed a preset threshold ε. We incrementally maintain each CB by taking into account the objects' anticipated movements. We capture the predicted update events (including split and merge events) of each CB during the continuous movement and process these events accordingly. At any time, clusters of different criteria can be constructed from the CBs, instead of the entire set of moving objects, which makes the construction processing cost efficient. Moreover, to reduce unnecessary computation of the network distance between the CBs, we adapt the network expansion method to combine CBs to construct the application-level clusters.

10.4.2 Construction and Maintenance of CBs

Initially, based on the CB definition, a set of CBs are created by traversing all edge segments in the network and their associated objects. The CBs are incrementally maintained after their creation. As time elapses, the distance between adjacent objects in a CB may exceed ε, and hence, we need to split the CB. A CB may also merge with adjacent CBs when they are within the distance of ε. Thus, for each CB, we predict the time when they may split or merge. The predicted split and merge events are then inserted into an event queue. Afterwards, when the first event in the queue takes place, we process it and update (compute) the split and merge events for affected CBs (new CBs if any). This process is continuously repeated. The key problems are: (1) how to predict split/merge time of a CB and (2) how to process a split/merge event of a CB.

The split of a CB may occur in two cases. The first is when a CB arrives at the end of the segment (i.e., an intersection node of the spatial network). When the moving objects in a CB reach an intersection node, the CB has to be split since they may head in different directions. Obviously, a split time is the time when the first object in the CB arrives at the node. In the second case, the split of a CB is when the distance between some neighboring objects moving on the segment exceeds ε. However, it is not easy to predict the split time since the neighborhood of objects changes over time. Therefore, the main task is to dynamically maintain the order of objects on the segment. We compute the earliest time instance when two adjacent objects in the CB meet as t_m. We then compare the maximum distance between each pair of adjacent objects with ε until t_m. If this distance exceeds ε at some time, the process stops and the earliest time exceeding ε is recorded as the split time of the CB. Otherwise, we update the order of objects starting from t_m and repeat the same process until some distance exceeds ε or one of the objects arrives at the end of the segment. When the velocity of an object changes over the segment, we need to re-predict the split and merge time of the CB.

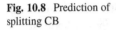

Fig. 10.8 Prediction of
splitting CB

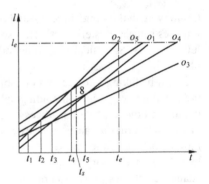

Figure 10.8 shows an example. Given $\varepsilon = 7$, we compute the split time as follows. At the initial time t_0, the CB is formed with a list of objects $<o_1, o_2, o_3, o_4, o_5>$. We first compute the time t_e when the first object (i.e., o_2) arrives at the end of the segment (i.e., l_e). For adjacent objects, we find that the earliest meeting time is t_1 at which o_2 and o_3 first meet. We then compare the maximum distances of each pair of adjacent objects during $[t_0, t_1]$ having no distance no larger than 7. At t_1, the object list is updated into $<o_1, o_3, o_2, o_4, o_5>$. In the same way, the next meeting time is at t_2 for o_2 and o_4. There are also no neighboring objects whose distance exceeds 7 during $[t_1, t_2]$. As the algorithm continues, at t_4, the object list becomes $<o_3, o_1, o_4, o_5, o_2>$ and t_5 is the next time for o_1 and o_4 to meet. When comparing neighboring objects during $[t_4, t_5]$, we find the o_4 and o_5 whose distance is longer than 7 at time t_s. Since $t_s < t_e$, we obtain t_s as the split time of the CB.

We now discuss how to handle a split event. If the split event occurs on the segment, we can simply split the CB into two ones and predict the split and merge events for each of them. If the split event occurs at the end of the segment, the processing would be more complex. One straightforward method is to handle the departure of the objects individually each time an object reaches the end of the segment. Obviously, the cost of this method is high. To reduce the processing cost, we propose a group split scheme. When the first object leaves the segment, we split the original CB into several new CBs according to objects' directions (which can be implied from *next_node*). On one hand, we compute a *to-be-expired time* (i.e., the time until the departure from the segment) for each object in the original CB and retain the CB until the last object leaves the segment. On the other hand, we attach a *to-be-valid time* (with the same value as *to-be-expired time*) for each object in the new CBs. Only valid objects will be counted in constructing application-level clusters. Figure 10.9 illustrates this split example. When CB_1 reaches J_1, objects p_1 and p_3 will move to the segment $<J_1, J_2>$, while p_2 and p_4 will follow $<J_1, J_6>$. Thus, CB_1 is split into two such that p_2 and p_4 join CB_3, and p_1 and p_3 form a new cluster CB_4. We still keep CB_1 until p_4 leaves $<J_4, J_1>$. As can be observed, the group split scheme reduces the number of split events and hence the cost of CB maintenance.

Fig. 10.9 Group split at an edge intersection. (**a**) When first object leaves. (**b**) When last object leaves

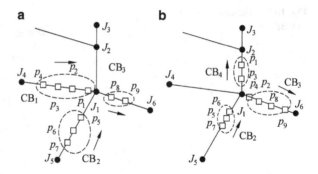

The merge of CBs may occur when adjacent CBs in a segment are moving together (i.e., their network distance $\leq \varepsilon$). To predict the initial merge time of CBs, we dynamically maintain the boundary objects of each CB and their validity time (the period when they are treated as the boundary of the CB) and compare the minimum distances between the boundary objects of two CBs with the threshold ε at their validity time. The boundary objects of CBs can be obtained by maintaining the order of objects while computing the split time. For the example in Fig. 10.8, the boundary objects of the CB are represented by (o_1, o_5) for validity time $[t_0, t_3]$, (o_3, o_5) for $[t_3, t_4]$, and (o_3, o_2) for $[t_4, t_e]$. The processing of the merge event is similar to the split event on the segment. We obtain the merge event and time from the event queue to merge the CBs into one CB and compute the split time and merge time of the merged CB. Finally, the corresponding affected CBs in the event queue are updated.

Besides the split and merge of a CB, new objects may come into the network or existing objects may leave. For a new object, we locate all CBs of the same segment that the object enters and see if the new object can join any CB according to the CB definition. If the object can join some CB, the CB's split and merge events are updated. If no such CBs are found, a new CB for the object is created and the merge event is computed. For a leaving object, we update its original CB's split and merge events if necessary.

10.4.3 CMON Construction with Different Criteria

This section discusses how to construct application-level clusters of different criteria from CBs. We focus our discussions on three common clustering criteria, i.e., distance based, density based, and K-partitioning.

10.4.3.1 Distance-Based CMON

A common clustering criterion is based on the minimum distance metric. The Minimum Distance CMON is defined as follows.

Fig. 10.10 The combination
of CBs

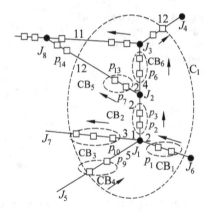

Definition 10.5. For each object in a Minimum Distance CMON (MD-CMON), the minimum network distance with other objects in the cluster is not longer than a user specified threshold δ ($\delta \geq \varepsilon$).

The requirement of $\varepsilon \leq \delta$ is necessary because it guarantees that a CB does not cross two clusters in the MD-CMON. The MD-CMON can be constructed by combining the CBs. Generally, for two CBs, we need to compute their network distance (i.e., the minimum network distance of their boundary objects) to determine whether to combine them. This simple method has a time complexity of $O(N^2)$, where N is the number of CBs. In order to reduce the computation cost, we adapt the incremental network expansion method to combine the CBs. The detailed algorithm can be found in Algorithm 24.

The algorithm starts with a CB and adds its adjacent nodes that are within δ to a queue Q using the Dijkstra algorithm. Take Fig. 10.10 as an example. Suppose $\delta = 10$ and the algorithm starts with CB_1. Thus, initially CB_1 is marked "visited" and J_1 is added to Q. The algorithm proceeds to dequeue the first node in Q (i.e., J_1). All adjacent edges of J_1 (except the checked edge $< J_6, J_1 >$) are examined. For each edge $< J_1, J_i >$, assuming $dist(J_1, J_i)$ to be the edge length, if J_i satisfies $dist(CB_1, J_1) + dist(J_1, J_i) \leq \delta$, J_i is added to Q and $dist(CB_1, J_i) = dist(CB_1, J_1) + dist(J_1, J_i)$. Moreover, all unvisited CBs on each adjacent edge are checked. For a CB_i on $< J_1, J_i >$, if $dist(CB_1, J_1) + dist(J_1, CB_i) \leq \delta$, CB_i is merged into CB_1's MD-CMON cluster. If $dist(CB_i, J_i) \leq \delta$ and J_i has not been added to Q, it is added to Q. The algorithm continues with the same process until Q becomes empty and the CBs around CB_1 are combined into a cluster C_1. Afterwards, the algorithm picks up another unvisited CB and repeats the same process until all CBs are visited.

10.4.3.2 Density-Based CMON

The second clustering criterion is Density based, which is suitable for filtering out noise data.

Algorithm 24: MD_CMON( )

---

foreach CB_i **do**
    if $CB_i.visited == false$ **then**
        Q = new priority queue;
        find edge n_x, n_y where CB_i lies;
        $CB = CB_i$; $C = CB$;
        $nextCB$ = Next CB on n_x, n_y from CB_i to n_y;
        while *(nextCB* $\neq$ *null) and Dist(CB.head,nextCB.tail)* $\leq \delta$ **do**
        Merge_Expand($CB,nextCB,C,n_x,n_y$);
        if *(nextCB* $==$ *null) and Dist(CB.head,n_y)* $\leq \delta$ **then**
            $B.node = n_y$; $B.dist = $ Dist($CB.head,n_y$);
            Enqueue(Q,B);
        while *notempty(Q)* **do**
            B = Dequeue(Q);
            foreach *node* n_z *adjacent to B.node* **do**
                $nextCB$ = Next CB from $B.node$ to n_z;
                if *(nextCB* $\neq$ *null) and Dist(B.node,nextCB.tail)+B.dist* $\leq \delta$ **then**
                    $newd_{n_z} = $ Dist($nextCB.head,n_z$);
                    Merge_Expand($CB,nextCB,C,B.node,n_z$);
                    while *(nextCB* $\neq$ *null) and Dist(CB.head,nextCB.tail)* $\leq \delta$ **do**
                        $newd_{n_z} = $ Dist($nextCB.head,n_z$);
                        Merge_Expand($CB,nextCB,C,B.node,n_z$);
                  end
                if *(no CBs on edge (B.node,n_z))* **then**
                    $newd_{n_z} = $ B.dist+Dist($B.node,n_z$);
                if *(nextCB* $==$ *null) and ($newd_{n_z} \leq \delta$)* **then**
                    $B_{new}.node = n_z$; $B_{new}.dist = newd_{n_z}$;
                    Enqueue(Q,B_{new});
             end
        end
end

Procedure MergeExpand($CB_1,CB_2,C,node_1,node_2$)

---

if $CB_2.visited == false$ **then**
    C =MergeClst(C,CB_2);
    $CB_1 = CB_2$; $CB_1.visited = true$;
    $CB_2 = $ Next CB from $node_1$ to $node_2$;
else
    C_1 =FindCluster($CB2$);
    C =MergeClst(C,C_1);
end

Definition 10.6. For each cluster in the density-based CMON (DB-CMON), the average density should be higher than a given threshold ρ. Moreover, there should not be any empty segment (without any objects lying on it) whose length is longer than E.

Suppose there are $m (m > 1)$ objects in a CB; the density of CB is $\frac{m}{\varepsilon(m-1)} > \frac{1}{\varepsilon}$. The second condition is necessary to avoid very skewed clusters. It is equivalent to the condition that for any object in the cluster, the nearest object is within a distance E. Thus, to construct the DB-CMON clusters from CBs, we require $\varepsilon \leq max\{E, \frac{1}{\rho}\}$.

The cluster formation algorithm is the same as the one described in Algorithm 24 except that the minimum distance constraint (transformed from the density constraint) is dynamic. Suppose the density of the current cluster with k objects is ρ' and a CB has m objects with a length of L. If a CB can be merged into the cluster, their minimum distance D must satisfy $\frac{k+m}{k/\rho'+L+D} \geq \rho$, i.e., $D \leq \frac{k+m+\rho(k/\rho'+L)}{\rho}$.

10.4.3.3 K-Partitioning CMON

K-Partitioning CMON is similar to the K-partitioning clustering method [13, 21]. It can be defined as follows.

Definition 10.7. Given a set of objects, K-partitioning CMON (KP-CMON) groups them into K clusters such that the sum of distances between all adjacent objects in each cluster is minimized.

According to the definition of CBs, the sum of distances between all adjacent objects in each CB is minimized. Therefore, it is intuitive to construct the KP-CMON clusters from the CBs. An exhaustive method is to iteratively combine the closest pairs of CBs until K clusters are obtained. This method requires to compute the distances between all pairs of CBs, which is costly. Hereby, we propose a low-complexity heuristic similar to the K-means algorithm [13, 21]. We initially select K CBs as the seeds for K clusters. For the remaining CBs, we assign them to their nearest clusters to minimize the sum of distances between adjacent objects. Note that this heuristic may not lead to the optimal solution. Suppose that in Fig. 10.11, the distances between CBs are: $dist(CB_2, CB_3) < dist(CB_2, CB_5) < dist(CB_3, CB_1) < dist(CB_2, CB_1) < dist(CB_3, CB_5)$ and that the initial seed CBs are CB_1 and CB_5 for $K = 2$. When CB_3 is checked, it will be assigned to the cluster of $\{CB_1\}$. Then, CB_2 will be assigned to the cluster of $\{CB_5\}$, which is different from the optimal solution where CB_2 and CB_3 should be grouped together since $dist(CB_2, CB_3) < dist(CB_2, CB_5)$. To compensate for such mistakes, we introduce the concept of Cross-CB. For adjacent CBs lying around the same node, if their minimum distance is less than ε, we group them into a Cross-CB. Then, the clustering algorithm is applied over the CBs and Cross-CBs.

Fig. 10.11 The cross-CB

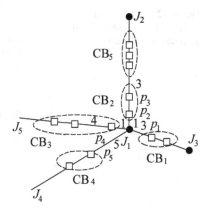

10.5 Clustering Trajectories Based on Partition-and-Group

Trajectory data of moving objects includes vehicle position data, hurricane track data, and animal movement data. While most existing trajectory clustering algorithms group similar trajectories as a whole, [14, 15] proposed a new *partition-and-group* framework for clustering trajectories and two types of clustering methods: *region based* and *trajectory based*.

First, each trajectory is partitioned into a set of trajectory partitions. Second, region-based clustering is performed recursively as long as homogeneous regions of reasonable size are found. The trajectory partitions that are not covered by homogeneous regions are passed to the next step. Third, trajectory-based clustering is performed repeatedly as long as discriminative clusters are found. The procedure of hierarchical region-based and trajectory-based clustering is shown in Fig. 10.12.

10.5.1 Partition-and-Group Framework

Partition-and-group framework partitions a trajectory into a set of line segments and then groups similar line segments together into a cluster. In particular, trajectory clustering based on this framework consists of the following two phases: (1) The partitioning phase: Each trajectory is optimally partitioned into a set of line segments. These line segments are provided to the next phase. (2) The grouping phase: Similar line segments are grouped into a cluster. Here, a density-based clustering method is exploited. The primary advantage of the partition-and-group framework is the discovery of common sub-trajectories from a trajectory database.

Based on the partition-and-group framework, a set of *clusters* $O = C_1, \ldots, C_{num_{clus}}$ as well as a *representative trajectory* for each cluster C_i can be

Fig. 10.12 The procedure of hierarchical region-based and trajectory-based clustering

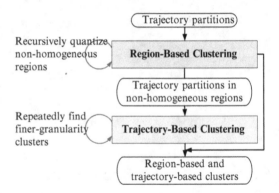

generated from a set of *trajectories* $I = TR_1, \ldots, TR_{num_{tra}}$, where the trajectory, cluster, and representative trajectory are defined as follows.

A *trajectory* is a sequence of multidimensional points. It is denoted as $TR_i = p_1 p_2 p_3 \ldots p_j \ldots p_{len_i} (1 \leq i \leq numtra)$. Here, $p_j (1 \leq j \leq len_i)$ is a d-dimensional point. The length len_i of a trajectory can be different from those of other trajectories. A trajectory $p_{c_1} p_{c_2} \ldots p_{c_k} (1 \leq c1 < c2 < \ldots < c_k \leq len_i)$ is called a *sub-trajectory* of TR_i.

A *cluster* is a set of trajectory partitions. A trajectory partition is a line segment $p_i p_j (i < j)$, where p_i and p_j are the points chosen from the same trajectory. Line segments that belong to the same cluster are close to each other according to the distance measure. Notice that a trajectory can belong to multiple clusters since a trajectory is partitioned into multiple line segments, and clustering is performed over these line segments.

A *representative trajectory* is a sequence of points just like an ordinary trajectory. It is an imaginary trajectory that indicates the major behavior of the trajectory partitions (i.e., line segments) that belong to the cluster. Notice that a representative trajectory indicates a common sub-trajectory.

Figure 10.13 shows the overall procedure of trajectory clustering in the partition-and-group framework. First, each trajectory is partitioned into a set of line segments. Second, line segments which are close to each other according to our distance measure are grouped together into a cluster. Then, a representative trajectory is generated for each cluster.

Algorithm 25 shows the skeleton of the above clustering process.

The distance function used in clustering line segments is composed of three components: (i) the perpendicular distance $(d_\perp)$, (ii) the parallel distance $(d_\parallel)$, and (iii) the angle distance (d_θ). They are illustrated in Fig. 10.14.

The *perpendicular distance* between L_i and L_j is defined as Formula (10.1). Suppose the projection points of the points s_j and e_j onto L_i are p_s and p_e, respectively. $l_{\perp 1}$ is the Euclidean distance between s_j and p_s; $l_{\perp 2}$ is that between e_j and p_e.

$$d_\perp(L_i, L_j) = \frac{l_{\perp 1}^2 + l_{\perp 2}^2}{l_{\perp 1} + l_{\perp 2}} \tag{10.1}$$

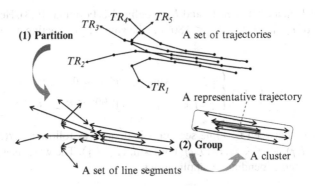

Fig. 10.13 An overall procedure of trajectory clustering in the partition-and-group framework

Fig. 10.14 Three
components of the distance
function for line segments

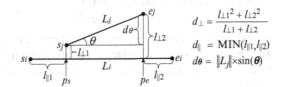

Algorithm 25: TRACLUS

---

input : A set of trajectories: $I = \{TR_1, \ldots, TR_{num_{tra}}\}$
output: A set of clusters: $O = \{C_1, \ldots, C_{num_{clus}}\}$, A set of representative trajectories
/* Partitioning Phase */
for $\forall TR \in I$ **do**
  Execute Approximate Trajectory Partitioning;
  Get a set L of line segments using the result;
  Accumulate L into a set D;
end
/* Grouping Phase */
Execute Line Segment Clustering for D;
Get a set O of clusters as the result;
for $\forall C \in O$ **do**
  Execute RepresentativeTrajectory Generation;
  Get a representative trajectory as the result;
end

The *parallel distance* between L_i and L_j is defined as Formula (10.2). Suppose
the projection points of the points s_j and e_j onto L_i are p_s and p_e, respectively. $l_{\|1}$
is the minimum of the Euclidean distances of p_s to s_i and e_i. Likewise, $l_{\|2}$ is the
minimum of the Euclidean distances of p_e to s_i and e_i.

$$d_\|(L_i, L_j) = MIN(l_{\|1}; l_{\|2}) \qquad (10.2)$$

The angle distance between L_i and L_j is defined as Formula (10.3). Here, $\| L_j \|$ is the length of Lj, and $\theta\,(0° \le \theta \le 180°)$ is the smaller intersecting angle between L_i and L_j.

$$d_\theta(L_i, L_j) = \begin{cases} \| L_j \| \times \sin(\theta), & \text{if } 0° \le \theta \le 90° \\ \| L_j \|, & \text{if } 90° \le \theta \le 180° \end{cases} \qquad (10.3)$$

The distance between two line segments is defined as follows: $dist(L_i, L_j) = w_\perp \times d_\perp(L_i, L_j) + w_\| \times d_\|(L_i, L_j) + w_\theta \times d_\theta(L_i, L_j)$. The weights $w_\perp, w_\|$, and w_θ are determined depending on applications.

10.5.2   Region-Based Cluster

A region-based cluster is formally defined through Definitions 10.8 and 10.9. This definition is very intuitive: a region-based cluster contains many of trajectory partitions of one major class, but very few of trajectory partitions of other minor classes.

Definition 10.8. A region in a 2-dimensional space is *homogeneous* if only one class c_{major} has trajectory partitions from $\ge \psi$ trajectories within the region, but all other classes do not. The class c_{major} is called the major class of the region, and other classes are called minor classes.

Definition above, ψ designates the *minimum population* of the major class in a homogeneous region. ψ typically shares the parameter value with *MinLns* to reduce the number of parameters to optimize. The two parameters, in fact, play the same role in region-based and trajectory-based clustering.

Definition 10.9. A *region-based cluster* is a set of trajectory partitions of the major class within a homogeneous rectangular region.

Figure 10.15 shows an example of region-based clustering. Suppose there is a set of trajectories from two classes c_1 and c_2, where the trajectories of c_1 are represented by solid lines and those of c_2 by dashed lines.

First, regions having one major (dominating) class are discovered as in (1). The regions B, F, and H are said to be homogeneous in the sense that they contain trajectories mostly of the same class. Second, the nonhomogeneous regions D and E are recursively quantized to find more of homogeneous regions. The region J is found to be homogeneous within E as in (2). These homogeneous regions are used as region-based clusters. Then, parts of trajectories in nonhomogeneous regions are passed to the next step.

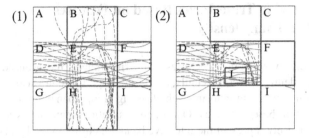

Fig. 10.15 An example of region-based clustering

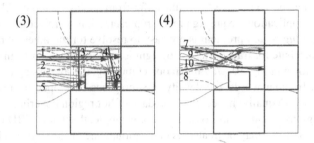

Fig. 10.16 An example of trajectory-based clustering

10.5.3 Trajectory-Based Cluster

A trajectory-based cluster is formally defined in Definition 10.10. This definition is essentially the same as the original definition of a trajectory cluster except the class constraint.

Definition 10.10. A *trajectory-based cluster* is a density-connected set of trajectory partitions of the same class.

Figure 10.16 shows an example of trajectory-based clustering. Suppose there is a set of trajectories from two classes c_1 and c_2, where the trajectories of c_1 are represented by solid lines and those of c_2 by dashed lines.

As shown in Fig. 10.16, common movement patterns of each class are discovered from nonhomogeneous regions as in (3). The patterns 3–6 are said to be discriminative in the sense that they are different from those of the other class. Then, the nondiscriminative patterns 1 and 2 are repeatedly investigated in finer granularity to find more of discriminative patterns. The horizontal movements are now represented by two patterns for each class rather than one. The patterns 7–10 newly discovered are discriminative as in (4). These discriminative patterns are used as trajectory-based clusters.

10.6   Clustering Trajectories Based on Features Other Than Density

Knowledge discovery from trajectories of moving objects has aroused a growing number of academic interests recently, due to the importance in various application scenarios such as traffic monitoring, vehicle navigation, urban planning, and various kinds of location-based services. One of the essential means takes advantage of density to concisely represent the intensive behaviors of moving objects, refine the density of spatial area as the number of distinct objects that pass through this area, and consequently to find out hot regions and more interesting results. Unfortunately, regions obtained based on density would suffer from the *big region* problem, which is critical for applications requiring precise representation of trajectory patterns.

In fact, the *big region* problem is not easy to resolve for the inherent reason that the density attribute for trajectories is not significant enough to constrain the size of region. As shown in Fig. 10.17, with an object moving, it may pass by a continuous spatial area. In other words, the density of subregions enveloped in this area will be added up by 1 equally; hence, the boundary of hot region covering neighboring dense subregions is really hard to restrict. For example, the whole CBD of a city or a not short motorway may be treated as one region, which is too loose to be applied to various environments, such as pattern-based movement prediction, as we would never know accurately where the object is once it enters this region, even when we can predict which region it goes next.

In this section, we introduce two novel metrics other than the density, together with corresponding evaluation functions and threshold determination strategies, and introduce a filter-refinement framework of hot region construction. In the filter step, we apply simple but efficient grid-based techniques to form dense regions. In the refinement step, the candidate regions are further reconstructed to obtain the compact regions.

10.6.1   Preliminary

This part studies the problem of discovering hot regions from trajectory databases first and then briefly introduces the framework and system. It starts by defining related concepts used in this work as follows.

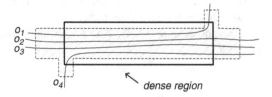

Fig. 10.17 The *big region* problem in density-based hot region clustering

Definition 10.11 (Region Density). Given a spatial region r and a time interval Δt, its *density* can be measured by $density(r, \Delta t) = N(r, \Delta t)/area(r)$, where $N(r, \Delta t)$ is the number of trajectories crossing region r during Δt and $area(r)$ is the area of r. $\square$

It has been proved that fixed shaped bounding containers are too rigid to be used, otherwise their restrictions on shape may make the hot regions suffer from various problems such as *redundant pattern* [17] or *flock-lossy* [9]. Correspondingly, density-based region discovering techniques have the advantages of both capturing clusters of arbitrary shapes and being robust with respect to noise.

Definition 10.12 (Hot Region). A region r is called a *hot region* if and only if it satisfies the following conditions:

- It is dense, i.e., its density is higher than a density threshold δ.
- It has a compact shape that it must be able to cover a disk with diameter α_1 and be covered by a disk with diameter α_2.
- No two hot regions overlap. $\square$

This definition gives us a coarse constraint on area and makes hot regions discovered more meaningful in practical, e.g., overcoming the phenomenon of *out-of-focus*. However, more precise arbitrary shape is recommended to use. Theoretically, a spatial region with a narrow and long belt shape may be assigned with high density, i.e., it satisfies the density constraint, but it has little meaning in practice. That is why we set a lower bound (i.e., α_1) for hot region.

Definition 10.13 (Hot Region Query). Given a trajectory database D, a density threshold δ, and α_1, α_2, find out all *hot regions* that satisfy the definition. $\square$

Any hot region in the query results is represented by a spatial region and a set of trajectories (intersecting this region) whose number divided by the area of the region is higher than a density threshold δ, and any region with diameter out of range $[\alpha_1, \alpha_2]$ will be filtered and refined.

To construct the hot regions that fulfill the definition above, a *filter-refinement* framework of hot region construction is proposed. In the *filter* step, we apply simple but efficient grid-based techniques to cluster dense regions, in the borrowed idea of the work [17] which is an optimization of the well-known DBSCAN [3] algorithm, and use the metric of density only to discover dense clusters. Detail of this part is roughly described as follows. We firstly partition the space objects moving on using a number of disjoint cells and consider the problem of finding the most dense cells. Since the movement of a moving object is known from geo-tags, we then extrapolate its trajectories and find all the cells that the object may cross. By maintaining the number of crossing per cell, we know each cell's density. We next reduce space by keeping only the cells that have been crossed so far and discard cells with zero density. With the constraints of the density threshold, we can finally combine neighboring cells into (coarse) dense regions with arbitrary shapes and sizes.

Likewise, regions obtained above would suffer from the *big region* problem. To solve this, in the *refinement* step, we introduce more rational metrics other than the

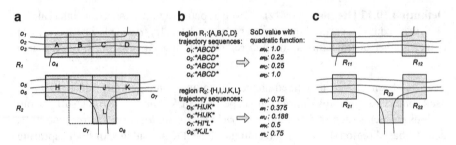

Fig. 10.18 An example of trend-based region reconstruction (TBRR). (**a**) Two big regions. (**b**) Score of domination (SoD). (**c**) Reconstruction ($minSoD = 0.5$)

density to reconstruct the regions whose sizes exceed that of the coverage constraints (α_1 and α_2), together with heuristic algorithms on how to reconstruct big regions and how to select proper values of the parameters used in the reconstruction algorithms. We proceed to detail these algorithms as follows.

10.6.2   Big Region Reconstruction

This part introduces two novel metrics and corresponding optimization algorithms to refine big regions as follows: *trend-based region reconstruction* (TBRR) and *dissimilarity-based region reconstruction* (DBRR), and then discusses how to select proper values for the parameters used in these two algorithms.

10.6.2.1   Trend-Based Region Reconstruction (TBRR)

It has been observed that, when an object enters (leaves) a region, the first (last) cells it meets potentially have greater effect on depicting the movement trend of this object than other cells of this region, because the contribution of inner cells is greatly dominated by entry-cells and exit-cells. Furthermore, cells visited frequently by moving objects are also much more important than those low-density ones.

Therefore, a novel weight metric, *score of domination* (SoD), is developed to weigh each cell according to its distance to the entry and exit where the trajectories pass through a big region. Next the total dominant score of each cell is calculated, and finally we can use these scores to reconstruct the big region into more compact pieces. Figure 10.18 illustrates an example of TBRR algorithm.

To count the distance between each cell and the nearest entry/exit of a given big region, each trajectory intersecting this big region is transformed into a sequence of the form $c_0 c_1 \ldots c_n$, where c_i is a spatial cell contained in this big region or the special character $*$, which indicates any other cells of the whole spatial universe. For instance, the sequence $o_7 : *HI * L*$ in Fig. 10.18b implies that the object o_7

stays outside region R_2 (it can be anywhere) at the beginning, then it is found in cell H, then it goes to region I, and after that it moves out of region R_2, until it is found in cell L and finally it exits the region again.

Let S_r denote a set of cell sequences, each of which is generated by a trajectory passing by the big region r, for any non-$*$ cell c on any sequence s, $s \in S_r$. The distance $d_{s,c}$ between cell c and nearest entry/exit (i.e., $*$) can be calculated with the following formula:

$$d_{s,c} = \begin{cases} min\{distance(c, *) \mid * \in s\} & \text{if } c \in s, \\ \infty & \text{otherwise.} \end{cases} \quad (10.4)$$

where $d_{s,c} \to \infty$ (i.e., $d_{s,c}^{-1} = 0$) if c_i never appears in the sequence, otherwise $1 \geq d_{s,c}^{-1} > 0$. To calculate SoD, we introduce three weight functions as below:

- Quadratic function: $\omega_c = \sum_{s \in S_r} d_{s,c}^{-2} / |S_r|$
- Exponential function: $\omega_c = \sum_{s \in S_r} 2^{1-d_{s,c}} / |S_r|$
- Factorial function: $\omega_c = \sum_{s \in S_r} d_{s,c}!^{-1} / |S_r|$

Figure 10.18c shows an example of the hot regions reconstructed with parameter $minSoD = 0.5$. The effectiveness of TBRR is sensitive to threshold $minSoD$. How to select a proper value for $minSoD$ is a complex problem; we discuss it later.

10.6.2.2 Dissimilarity-Based Region Reconstruction (DBRR)

Another interesting observation is that the cells passed by diverse dissimilar trajectories should be paid more attentions. This is because the differentiation reflects the more precise distribution of objects.

Based on this, we have another novel weight metric, *degree of dissimilarity* (DoD), which measures the overall dissimilarity of a given cell. Before calculating the degree of dissimilarity, we formalize the concepts (and corresponding variables) used in weight functions as follows.

Neighborhood: Given a cell c, we call the cells in immediate proximity to cell c as its neighbors (at most eight). We term the set of c's neighbors as NB_c, together with the neighboring cells of region r as NB_r, where $NB_r = \{c \mid \exists c_i \in r \land c \in NB_{c_i} \land c \notin r\}$.

Cell bucket: For each cell c, we term the set of objects passing by cell c as $H(c)$ (i.e., $N_c = |H(c)|$). Figure 10.19b shows an example of these cell buckets, parts of which are gray colored corresponding to external neighboring cells.

Coverage region: We extend a big region r with its neighboring cells and get a virtual coverage region $\bar{r}$. And we further remove the empty cells from NB_c and NB_r to reduce the scale of dissimilarity calculation and then get a compact neighbor set NB_c' and a compact coverage region $\bar{r} = r \cup \{c \mid c \in NB_r \land |H(c)| > 0\}$. Figure 10.19a illustrates an example of coverage regions, $\overline{R_1}$ and $\overline{R_2}$, in which dashed squares denote those external neighboring cells.

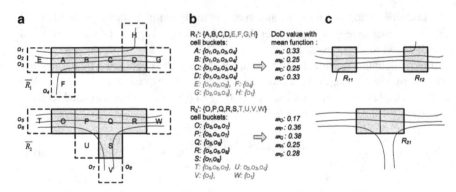

Fig. 10.19 An example of dissimilarity-based region reconstruction (DBRR). (**a**) Two extended big regions. (**b**) Degree of dissimilarity (DoD). (**c**) Reconstruction ($minDoD = 0.3$)

With variables above, given a big region r and two neighboring cells c_i, c_j which are both contained in coverage region $\bar{r}$, the *dissimilarity* between c_i and c_j is measured by

$$diss(c_i, c_j) = 1 - \left| \frac{H(c_i) \cap H(c_j)}{H(c_i) \cup H(c_j)} \right| \tag{10.5}$$

The dissimilarity of cell c ($c \in r$) is a summary of the dissimilarly between c and its neighboring cells, i.e.,

$$DISS(c) = \sum_{c_i \in \bar{r}} diss(c, c_i). \tag{10.6}$$

In addition, we define the relative *hotness* of cell c as follows:

$$hotness(c) = \frac{|H(c)|}{|\bigcup_{c_i \in r} H(c_i)|}. \tag{10.7}$$

Next, the standard degree of dissimilarity of cell c ($c \in r$) can be calculated by the following aggregate functions:

- Square root of 2nd-order origin moment: $\omega_c = \sqrt{\frac{\sum_{c_i \in \bar{r}} diss(c, c_i)^2}{|\bar{r}|}} \cdot hotness(c)$
- Summary value estimation function: $\omega_c = \frac{DISS(c)}{max\{DISS(c_i) \mid c_i \in r\}} \cdot hotness(c)$
- Mean value estimation function: $\omega_c = \frac{\sum_{c_i \in NB_c'} diss(c, c_i)}{|\bar{r}|} \cdot hotness(c)$

Figure 10.19 demonstrates the whole process of DBRR. Likewise, the effectiveness of DBRR is sensitive to threshold $minDoD$.

10.6.3 Parameters Determination in Region Refinement

Proper values of *minSoD* and *minDoD* are difficult to find in some applications since they are dependent on the characteristics of trajectory data; normally little a priori knowledge of the data can be used to enlighten users on parameter selection. This part introduces the guidelines for determining these parameters. To reconstruct big regions adequately and consequently and fasten the execution of refinement, it needs an efficient method to determine the individual parameters for each big region during the reconstruction algorithms.

First, a *recursive cut-and-try method* (RCTM) is proposed to refine the coarse regions with individual parameters. Given a big region r consisting of n cells and assuming each cell has annotated with a weight (i.e., SoD or DoD), RCTM is carried out as follows:

1. Sort cells in r according to their weights in descending order.
2. Choose cell c_i with *heuristic strategies* from the sequence of sorted cells.
3. Reconstruct r using c_i's weight as the individual parameter.
4. If big regions still exist, reconstruct each of them from step 1.
5. Till r is refined adequately.

It is easy to prove that the RCTM method converges. Each round of reconstruction (i.e., step 3 of RCTM) discards at least $i - 1$ cells from r, and the overall area of regions after reconstruction is smaller than that of r. After finite iterations of reconstruction, a (or empty) set of appropriate regions are obtained.

Next, introduce two *heuristic strategies* to determine which cell to choose:

- With the first strategy, we choose the cell c_i closest to $p\%$ position in the sequence S, e.g., the middle of the sequence $c_m, m = \lceil 50\% \times S.len \rceil$.
- With the second strategy, we find the largest variance between two adjacent weights of cells in S and then select the cell with the smaller weight.

The first strategy is quantitative, and it can always reduce the size of big regions effectively. Correspondingly, the second strategy is qualitative, as it considers the nonuniformity of cell weight distribution. The idea behind the second strategy is to find a relatively small threshold that achieves a reasonable effectiveness of reconstruction. The principle that both have in common is they are insensitive to the size of regions. In other words, no matter how big the region is, algorithms with these parameters can reconstruct the coarse regions just as effectively.

10.7 Summary

In this chapter, we studied the problem of clustering moving objects in a spatial network and proposed a framework to address this problem. By introducing a notion of cluster block, this framework, on one hand, amortizes the cost of clustering into

CB maintenance and combination based on the object movement feature in the road network; on the other hand, it efficiently supports different clustering criteria. We have exploited the features of the road network to predict the split and merge of CBs accurately and efficiently. Three different clustering criteria have been defined, and the cluster construction algorithms based on CBs were proposed.

References

1. Agrawal R, Gehrke J, Gunopulos D, Raghavan P (1998) Automatic subspace clustering of high dimensional data for data mining applications. In: Proceedings of the ACM SIGMOD international conference on management of data (SIGMOD 1998), Seattle, pp 94–105
2. Ankerst M, Breunig MM, Kriegel HP, Sander J (1999) OPTICS: ordering points to identify the clustering structure. In: Proceedings of the ACM SIGMOD international conference on management of data (SIGMOD 1999), Philadelphia, pp 49–60
3. Ester M, Kriegel H, Sander J, Xu X (1996) A density-based algorithm for discovering clusters in large spatial databases with noise. In: Proceedings of the 2nd ACM SIGKDD international conference on knowledge discovery and data mining (KDD 1996), Portland, pp 226–231
4. Fisher D (1987) Knowledge acquisition via incremental conceptual clustering. Mach Learn 2:139–172
5. Giannotti F, Nanni M, Pinelli F, Pedreschi D (2007) Trajectory pattern mining. In: Proceedings of the 13th ACM SIGKDD international conference on knowledge discovery and data mining (KDD 2007), San Jose, pp 330–339
6. Guha S, Rastogi R, Shim K (1998) CURE: an effcient clustering algorithm for large databases. In: Proceedings of the ACM SIGMOD international conference on management of data (SIGMOD 1998), Seattle, pp 73–84
7. Han J, Kamber M (2005) Data mining: concepts and techniques, 2nd edn. Morgan Kaufmann Publishers Inc., San Francisco
8. Jain AK, Dubes RC (1988) Algorithms for clustering data. Prentice Hall, Englewood Cliffs
9. Jeung H, Yiu HL, Zhou X, Jensen CS, Shen H (2008) Discovery of convoys in trajectory databases. In: Proceedings of the 34th international conference on very large data bases (VLDB 2008), Auckland, pp 1068–1080
10. Jin W, Jiang Y, Qian W, Tung AKH (2006) Mining outliers in spatial networks. In: Proceedings of the 11th international conference on database systems for advanced applications (DASFAA 2006), Singapore, pp 156–170
11. Kalnis P, Mamoulis N, Bakiras S (2005) On discovering moving clusters in spatio-temporal data. In: Proceedings of the 9th symposium on spatial and temporal databases (SSTD 2005), Angra dos Reis, pp 364–381
12. Karypis G, Han EH, Kumar V (1999) Chameleon: hierarchical clustering using dynamic modeling. IEEE Comput 32(8):68–75
13. Kaufman L, Rousseeuw PJ (1990) Finding groups in data: an introduction to cluster analysis. Wiley, New York
14. Lee JG, Han J, Li X, Gonzalez H (2008) TraClass: trajectory classification using hierarchical region-based and trajectory-based clustering. In: Proceedings of VLDB 2008, Auckland, pp 24–30
15. Lee JG, Han J, Whang KY (2007) Trajectory clustering: a partition-and-group framework. In: Proceedings of the 2007 ACM SIGMOD international conference on management of data (SIGMOD 2007), Beijing, pp 593–604
16. Li YF, Han JW, Yang J (2004) Clustering moving objects. In: Proceedings of the 10th ACM SIGKDD international conference on knowledge discovery and data mining (KDD 2004), Seattle, pp 617–622

17. Mamoulis N, Cao H, Kollios G, Hadjieleftheriou M, Tao Y, Cheung D (2004) Mining, indexing, and querying historical spatiotemporal data. In: Proceedings of the 10th ACM SIGKDD international conference on knowledge discovery and data mining (KDD 2004), Seattle, pp 236–245

18. Martin E, Kriegel HP, Sander J, Xu X (1996) A density-based algorithm for discovering clusters in large spatial databases with noise. In: Proceedings of the 2nd ACM SIGKDD international conference on knowledge discovery and data mining (SIGKDD 1996), Portland, pp 226–231

19. Nanopoulos A, Theodoridis Y, Manolopoulos Y (2001) C2P: clustering based on closest pairs. In: Proceedings of the 27th international conference on very large data bases (VLDB 2001), Roma, pp 331–340

20. Nehme RV, Rundensteiner EA (2006) SCUBA: scalable cluster-based algorithm for evaluating continuous spatio-temporal queries on moving objects. In: Proceedings of the 10th international conference on extending database technology (EDBT 2006), Munich, pp 1001–1019

21. Ng RT, Han J (1994) Efficient and effective clustering methods for spatial data mining. In: Proceedings of the 20th international conference on very large data bases (VLDB 1994), Santiago de Chile, pp 144–155

22. Wang W, Yang J, Muntz R (1997) STING: a statistical information grid approach to spatial data mining. In: Proceedings of the 23rd international conference on very large data bases (VLDB 1997), Athens, pp 186–195

23. Yiu ML, Mamoulis N (2004) Clustering objects on a spatial network. In: Proceedings of the ACM SIGMOD international conference on management of data (SIGMOD 2004), Paris, pp 443–454

24. Zahn C (1971) Graph-theoretical methods for detecting and describing gestalt clusters. IEEE Trans Comput 20(1):68–86

25. Zhang Q, Lin X (2004) Clustering moving objects for spatio-temporal selectivity estimation. In: Proceedings of the 15th Australasian database conference (ADC 2004), Dunedin, pp 123–130

26. Zhang T, Ramakrishnan R, Livny M (1996) BIRCH: an effcient data clustering method for very large databases. In: Proceedings of the ACM SIGMOD international conference on management of data (SIGMOD 1996), Montreal, pp 103–114

Chapter 11
Dynamic Transportation Navigation

Abstract The widespread use of GPS navigation and trip planning on web has aroused considerable interests in fast and scalable path query processing. Existing research has mainly focused on static route optimization where the traffic network is assumed to be stable. Nevertheless, in most cases, route planning is in presence of frequent updates to the traffic graph due to the dynamic nature of traffic network, and such updates always greatly affect the performance of route planning. Most existing methods, however, cannot efficiently support traffic aware route planning. In this chapter, we overview some existing approaches for dynamic transportation navigation, and then introduce an novel traffic aware route planning strategy, in which a set of effective techniques are employed to avoid both unnecessary calculations on huge graph and excessive re-calculations caused by traffic condition updates.

11.1 Introduction

As the prices of equipment like smart cell phones, PDA devices, wireless modems, and GPS devices continue to drop rapidly, the number of wireless subscribers worldwide will soar. As a result, location based services is growing in popularity in recent years. Many online route planning services such as Google Maps and Microsoft MapPoint have become one of the most important tools for our life nowadays. In addition, the popular use of location based services such as GPS navigation and logistic control has led to great interests in real time route planning techniques.

Finding the shortest or fastest paths from road network is a classical problem that has been intensively studied. Most route search algorithms [1, 17, 22] mainly focus on static network, where road conditions are assumed to be stable. But in reality, a key feature of actual road condition is its high dynamics: just consider how road speed changes in peak hours. As such, it is essential to achieve traffic aware navigation, so that users can be guided to bypass congestions and continuously follow the best path, without been affected by new congestions occurred ahead.

The major challenge of traffic aware navigation comes from the expensive computational overhead caused by frequent road condition updates and the scalable graph of road network. On the Boston road network that contains over 40,000 links [19], even a single calculation for fastest path search could cost iCarTel (an iphone application) several seconds. It is obviously not realistic to simply re-plan the route for each road condition update, and it is thus crucial to improve classical algorithms, by finding ways to reduce the search space of each route search and to avoid time of re-computation caused by road condition changes.

Finding the shortest paths from road network is a classical problem, with many efforts have been made on it so far. Dijkstra algorithm is a well known algorithm that finds shortest path on graph, and A* and its variation [15, 18] improves the search performance by using effective heuristics. Later, some pre-computation based techniques like estimation [21], transit node routing [1], network indexing [2, 22, 23, 27], hierarchical routing [8, 24] and landmark [9] are further proposed to speed up route search. Also, some other major efforts related to this topic mainly include probabilistic path queries [12], dynamic kNN [20], path oracles and efficient processing [6, 7, 25], skyline queries [4, 16], trip planning with multiple destinations [17] and complex road network structure [13].

Above pure distance-dependent approaches are not effective because they do not consider the road condition factor in route planning. Recently, increasing attentions have been put on time-dependent shortest path search problem, where the shortest path search is based on a dynamic graph due to speed changes. Classical algorithms like D* can be used to find shortest path continuously on dynamic graph. Also some mining based approaches are proposed: Gonzalez et al. in [10] use mining techniques to derive frequent driving patterns, and then compute the adaptive fastest path based on the match pattern. Route update is made when the matched pattern switches. Ding et al. in [5] proposed Dijkastra based algorithm proposed to find the best departure time and fastest path over the network, according to the speed pattern based on statistical average road condition in different time intervals. Also, a critical-time-point approach is presented in [11] to generate best routes and their corresponding time intervals, and the issue of finding fastest paths on a road network with speed patterns is discussed in [14]. However, they only rely on the speed patterns derived from traffic data, which means, the searched path is un-likely to be optimal because the real time information on road condition is not considered.

More recently, some novel route planning approaches with considering real-time road condition are proposed. Malviya et al. in [19] targeted to answer continuous route planning queries over a road network in presence of speed updates on road segments. Its basic idea is to calculate k fastest paths (based on speed patterns) between any two vertexes with variance guarantee at the build time, and then to keep ranking the k fastest paths from current position to destination at the run time. Also, a heuristic based bidirectional route planning algorithm is proposed in [3] to speed up the search process. However, above approaches requires huge computational overhead, and they are thus not feasible for applications where the road conditions are frequently updated.

In this chapter, we will review some typical existing strategies that can be used for dynamic transportation navigation. Afterwards, some novel model and algorithms are particularly introduced to support traffic aware route planning.

11.2 Typical Dynamic Transportation Navigation Strategies

The task of dynamic transportation navigation has received considerable attention in the research literature. Existing strategies mainly focus on path search and update policies. In this section, we introduce some typical methods including the D*, Hierarchy Aggregation Tree (HAT) based navigation.

11.2.1 D* Algorithm

D* [26] is an algorithm capable of planning paths in unknown, partially known and changing environments in an efficient, optimal and complete manner. The name of the D* was chosen because it resembles A*, except that it is dynamic in the sense that arc cost parameters can change during the problem solving process.

Like Dijkstra and A* algorithm, D* maintains the "OPEN list" to propagate information about changes to the arc cost function and to calculate path costs to states in the space. Each state X has an associated tag $t(X)$ having one of several states:

- NEW: it has never been placed on the OPEN list
- OPEN: it is currently on the OPEN list
- CLOSED: it is no longer on the OPEN list
- RAISE: its cost is higher than the last time it was on the OPEN list
- LOWER: its cost is lower than the last time it was on the OPEN list

In the D* algorithm, a node is iteratively selected from the OPEN list from the OPEN list according to certain criterions. Then the weight changes to all neighboring nodes caused by this node are calculated and corresponding updates are made on OPEN list. This process is termed "expansion". In contrast to A*, the search process of D* is in backwards direction from destination. Each expanded node has a back pointer which refers to the next node leading to the target, and each node knows the exact cost to the target. The algorithm finishes when the start node is the next node to be expanded, and the route to the destination can be found by simply following the back pointers.

When an obstruction occurs on the planned path, all the affected points (by obstruction) are inserted to the OPEN list, and this list is then marked as RAISE list. For a node in RAISE increases cost, the algorithm checks its neighbors and examines whether it can reduce the node's cost. If not, the RAISE state is propagated to all descendants of the nodes that have back pointers to them. These nodes are evaluated, and the RAISE state passed on, forming a wave. When a RAISED node can be reduced, we update its back pointer and pass the LOWER state to its neighbors. By this point, a whole series of other points are prevented from being 'touched' by the wave with the help of threshold. The algorithm has therefore only worked on the points which are affected by change of cost.

11.2.2 Hierarchy Aggregation Tree Based Navigation

Since classical spatial indices like R-tree and Quad-tree are not suitable for optimal path searching in traffic environments, a new indexing method named Hierarchy Aggregation Tree (HAT) can be used to improve the efficiency of query processing. It is based on two structures: road and region (similar to MBR). In addition, it contains supplement information that stores an aggregated value over it. The principal functionality of aggregated information is to filter the regions having a high traffic density on the same hierarchy level.

HAT is set up based on the spatial information of roads. Unlike the R-tree, HAT references edges and nodes so that it avoids dead space. The search is more efficient because node MBRs do not overlap. This method is inspired from the non-overlapping index like Quad-tree, except that the space partitioning in HAT may be skewed and the resulting tree will be balanced. For each region, HAT stores traffic density information at different granularity levels to provide a filter capability for the path-finding process.

HAT is constructed by partitioning the index space recursively. The space is divided according to the distribution of network segments, by an adaptive and recursive split of space in four sub-regions. When the amount of roads, namely capacity in a leaf node N, exceeds a predefined threshold B for split, N is to be split and the corresponding region to be partitioned into four sub-regions. The split of HAT satisfies the following two rules: (1) Capacities in four sub-regions should almost be the same; (2) The sub-regions crossed by a road should be as few as possible, namely, the copies of entries for the road should be as few as possible.

Adopting the aforementioned index structure and navigation method, an intelligent city traffic control system have been designed and implemented, named DyNSA (Dynamic Navigation System based on moving objects stream Aggregation), with the aim of providing high quality of dynamic navigation services. An overview of this system architecture is shown in Fig. 11.1.

This system consists of multiple managers: Traffic Information Receiver (TIR), Traffic Information Manager (TIM) and Query Processor. TIR is an information receiver, which continuously sends traffic information to TIM. In TIM, aggregated

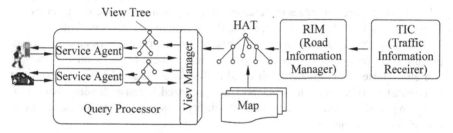

Fig. 11.1 System architecture of DyNSA

information of each road segment is refreshed regularly according to current traffic information and the region aggregation on each HAT's hierarchy level is thus recalculated. Query process is in charge of users' navigation requests. When a navigation request arrives, it is sent to a View Manager, and then a corresponding view tree on HAT will be created. The Service Agent will perform the view-based hierarchy search on it, and finally, the optimal path will return to the user. Since the View Manager maintains a consistency between the view tree and the HAT, a recalculation will happen if necessary and will be sent to the user until she/he arrives at her/his destination. The underlying index structure of RIM and Query Processor are both based on HAT.

11.3 Incremental Route Search Strategy

This section introduces an incremental route planning approach proposed in [28] to achieve efficient traffic aware route planning. Particularly, each time we compute a partial path rather than the whole route from start to end. It enables the computational overhead to be greatly reduced for two main reasons: firstly, excessive re-calculations (particularly on faraway road segments) due to frequent road condition updates can be avoided; secondly, it guarantees the real time response to route queries because each partial route search is restricted in a small region. In our approach, a set of graph reduction and filtering mechanisms are used to improve algorithm efficiency, and issues like driving flexibility and congestion evolution are considered to improve the effectiveness of route planning under dynamic road networks.

11.3.1 Problem Definitions

Given a road network defined as a directed graph $G = \langle V, E \rangle$, where $V = \{v\}$ is the set of vertices representing road ends or intersections, and $E = \{(v_i, v_j) \mid v_i, v_j \in$

V} is the set of directed edges representing road segments. Assume v is a vertex on G, we use $ind(v)$ and $outd(v)$ to represent the in-degree and out-degree of v.

The in-degree and out-degree of vertices are important attributes for route planning. Given a vertex v, congestions are more likely to occur on v if $ind(v)$ is high because of the greater in-flux traffic flow, and we should thus avoid them. In contrast, vertices with high $outd(v)$ are preferred because drivers have more (outgoing) paths to choose from. Such flexibility is very useful to bypass the new congestion occurs ahead.

Definition 11.1 (Road Condition). The road condition of a road network can be expressed by $C = <S, T, D>$, where S and T are the speed and required traveling time for road segments, and D denotes the time duration when this road condition is valid. For each road segment e, the speed on e is represented as $s_e \in S$, and the time required for passing e is $t_e \in T$. Given a time t, we use function $traffCond(t)$ to return the road condition C which satisfies $t \in C.D$.

Definition 11.2 (Fastest Route Query). In a dynamic road network G, a fastest route query is defined as $qry = \langle src, dst, CS \rangle$, where src is the source vertex and dst is the destination vertex specified by users, and CS is a set of road conditions that affect route planning (having temporal intersection with the travel).

Problem Definition. We target to solve the problem of traffic aware route planning which is formally defined as: Given a fastest route query $qry = \langle src, dst, CS \rangle$ on road network $G = \langle V, E \rangle$, particularly with traffic condition updates $CS = \langle C0, C1, C2, \ldots \rangle$ we process the query for a continuous optimal path (route) $pth = (v_s, v, v', \ldots, v_t)$ on this dynamic road network that satisfies the following spatio-temporal optimization goals and constraints:

1. Spatial constraints: source vertex $v_s = src$ and end vertex $v_t = dst$;
2. Traffic condition constraints: The planned route pth must be temporally consistent with traffic conditions CS;
3. Optimization goal: The total travel time should be minimized.

It is a computationally hard problem due to the huge scale of graph and the continuous recalculation caused by the excessive traffic condition updates on this graph. However, GPS navigation requires efficient query processing for immediate response. Therefore, an efficient route search approach is highly sought after.

Targeting to this problem, we propose a novel strategy for traffic aware route planning. Firstly, we apply basic graph reduction to facilitate route planning. Then we select the top-k intermediate destinations according to a set of spatio-temporal criterions. We afterwards introduce how to plan the route based on the top-k intermediate destinations. In this way, we can support real time response and avoid excessive re-calculation due to frequent road condition updates. Lastly, we discuss the monitoring technique to achieve adaptive route planning.

Fig. 11.2 Basic graph
reduction

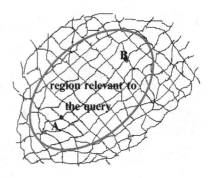

11.3.2   Pre-computation

In initialization phase, we conduct a basic graph reduction to settle a small sub-graph as the region relative to route query processing: an eclipse region G' is efficiently derived from the whole space G in the same way as [19], as shown in Fig. 11.2.

Since an ellipse is the simplest geometric shape that we can employ besides a circle to deal with distance, we first introduce how to reduce the search space using some features of an ellipse. An ellipse is the trajectory of a point whose combined distance to two foci is fixed, and the distance is equal to the length of its principle axis. An important characteristic of an ellipse is that all points within the ellipse are closer to the two foci than those on its boundary, and all points outside an ellipse to the two foci are farther than those on its boundary.

In our approach, we first calculate the network distance d between source vertex *src* and destination *dst*. It is simply processed by A* algorithm because it is effective for distance based search. Then we use d as the length of the principle axis and use the locations of the two vertexes *src* and *dst* to position the foci to construct an ellipse. It can be proved that if there is a shortest path between *src* and *dst*, then this path lies within the ellipse: Assume a vertex v belongs to the shortest path of *src* and *dst*, and that v is located outside the ellipse; we then have $|src \rightarrow v| + |v \rightarrow dst| > d$. As a result, even if there exist straight-line paths between *src* and v as well as v and *dst*, the length of this path must be greater than d. Therefore, vertex v does not belong to the shortest path between *src* and *dst*. Based on this theorem, we can use an ellipse to prune the vertexes that cannot possibly be on the shortest path.

Therefore, only the road segments in this ellipse region are considered as relevant to route planning. However, as only positions of source and end vertexes are considered, this eclipse region is very likely to be over-sized, and we only use it to set a base for route planning operations used in remaining sections.

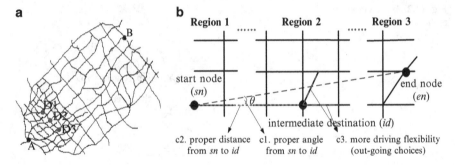

Fig. 11.3 (**a**) Intermediate destinations. (**b**) Evaluation criterions

11.3.3 Top-K Intermediate Destinations

In traditional approaches, a path strictly from source point to destination is usually planned in each time. However it is not effective for earliest arriving route search because the frequent updates on road condition are likely to cause excessive re-calculation, particularly on faraway road segments. For example, if congestion occurs on a road segment that is part of planned route, re-planning is needed to guarantee service quality. For efficiency purpose, it is thus reasonable to plan partial path in a limited scope, rather than plan the whole route. Re-calculations caused by dynamic road condition can be greatly reduced accordingly. To set the boundary of route search properly, we must select some intermediate destinations as shown in Fig. 11.3a, toward which effective route planning are conducted afterwards.

The selection of intermediate destinations must follow a set of spatio-temporal standards. First of all, the direction from source point to intermediate destination has great evaluating merit. To evaluate the direction preference of selecting v_x as intermediate destination, we use $dw(v_x)$ to measure the direction weight of v_x as:

$$dw(v_x) = cos(\overrightarrow{v_s, v_x}, \overrightarrow{v_s, v_d})$$

where $cos(\overrightarrow{v_s, v_x}, \overrightarrow{v_s, v_d})$ is the cosine value of an angle regarding to two lines $\overrightarrow{v_s, v_x}$ and $\overrightarrow{v_s, v_d}$, e.g., "c1" in Fig. 11.3b. It can be seen as the difference between direction to intermediate destination and that to final destination. We judge their direction are consistent if their angle is small, so we prefer vertexes with less value of $dw(v_x)$ because the cosine value is in reverse proportion to the degree of angle.

Meanwhile, the position of a vertex regarding to source and end points is an important criterion according to "c2" of Fig. 11.3b. A faraway intermediate destination may cause excessive re-calculation due to the frequent updates on traffic condition. Also, the distance should not be too close because the global view is neglected: it is hard to satisfy global optimization when searching a partial path to an intersection 200 m away. To achieve a good balance between reducing re-calculation (not too faraway) and achieving global optimization (not too close), we use position weight $pw(v_x)$ to measure intersection v_x as:

$$pw(v_x) = score - |dis_{EU}(v_s, v_x) - dis_{best}|$$

where $dis_{EU}(v_s, v_x)$ denotes the Euclidean distance (e.g., 10 miles) from v_s to v_x, and dis_{best} is the best distance from v_s to intermediate destination based on statistics. $score$ is the standard weight of each vertex (for been selected), and distance $|dis_{EU}(v_s, v_x) - dis_{best}|$ can be seen as the penalty on vertex v_x's position.

Another important criterion to evaluate intermediate destinations is the flexibility of future driving. According to 'c3' of Fig. 11.3b, high flexibility means better capability for exception handling, e.g., to bypass new occurred congestions. Driving flexibility of an intermediate destination is determined by a set of spatio-temporal features. It is obvious that more out-going paths from an intersection give us more flexibility to choose. Among out-going edges, those in the same direction to final destination are definitely preferred. Wrapping up these issues, the flexibility weight $fw(v_x)$ for selecting vertex v_x as intermediate destination is calculated as:

$$fw(v_x) = \sum_{v \in FV(v_x)} cos(\overrightarrow{v_x, v}, \overrightarrow{v_x, v_d})$$

where $FV(v_x) = \{v | e = (v_x, v) \in E\}$ is the forward vertices set of v_x. More outgoing edges give drivers greater flexibility of route selection. For each out-going edge to v from v_x, the angle between $\overrightarrow{v_x, v}$ and $\overrightarrow{v_x, v_d}$ is preferred to be small because it is in the consistent direction to destination. We thus use the sum of cosine value to evaluate the preference of its out-going paths.

To select proper intermediate destinations, issues mentioned above like spatial features and driving flexibility must be considered, as they benefit us to reduce re-calculations and to be more reliable under dynamics. In particular, the selectivity of intermediate destinations follows the following criterion:

$$w(v_x) = dw(v_x)/cd + pw(v_x)/cp + fw(v_x)/cf.$$

Where $dw(v_x)$ is the direction weight, $pw(v_x)$ is the position weight, $fw(v_x)$ is the flexibility weight. Factors cd, cp and cf represent the standard direction weight, position weight and flexibility weight required for capable intermediate destinations respectively. Thus, only vertices that satisfies $w(v) \geq 3$ are suitable as intermediate destination. By ranking $w(v)$ on vertices meeting this requirement and not belonged to the congested region, k best vertices and final destination are selected in intermediate destination candidate set IDC. Route search is then conducted towards vertices in IDC. If intermediate destination candidate cannot be detected, A* algorithm is simply used to search time-dependent shortest path to final destination v_t.

11.3.4   Route Search and Update

Route search is made to find a path (possibly a partial route) to one of the top-k intermediate destination selected from last step. We propose a novel algorithm to search the partial route efficiently. The exact time from the source vertex to any intermediate destination is considered, while from the intermediate destination to final destination we use the estimated time to avoid traversing huge scale faraway road segments. Notice that it also contributes to reduce re-calculations caused by dynamics of road conditions on those road segments.

Compared with the conventional shortest path search, the targets of the route search in our strategy are intermediate destinations here, rather than the final destination. Therefore, we present a partial path search algorithm, which can efficiently find the partial path to an intermediate destination in terms of minimal estimated time cost from static of view (current road condition).

Algorithm 26: Partial path search algorithm (src, dst, G', IDC)

---

input : src, dst, G' and IDC are the source vertex, destination, search graph after filtering and
            set of intermediate destination candidates respectively

output: *ppath* is the optimal partial path to return

Create the set of reached vertexes $RV = \{src\}$ and OPEN list $OL = \{$vertexes have an
edge from src to it$\}$;

while *no intermediate destination in IDC is reached* **do**

    if *OL is empty* **then**

        stop and report error;

        /** destination cannot be reached*/

    end

    $MinDistIncr = 0$;

    $toExtendedVert$ = null ;

    for *vertex* $v \in OL$ **do**

        compute estimated time D via v to dst, D is the sum of actual time from src to v

        and estimated time from v to dst based on EU distance and allowed speed ;

        if $D < MinDistIncr$ **then**

            $MinDistIncr = D$;

            $toExtendedVert = v$;

        end

    end

    set v as reached, and record the shortest path to it ;

    if $v \in IDC$ **then**

        set partial path *ppath* = as the recorded shortest path to v ;

        break ;

    end

    Update the OPEL list OL regarding to vertex v ;

end

return *ppath* ;

Algorithm 26 specifies how the route search algorithm works. Like most shortest path search algorithms, we first create and initialize the set of reached vertexes RV and OPEN list OL. RV contains all vertexes we have processed such that the path from current position to them are known, and OL contains all visible vertexes open for next expansion. In each step, we expand on the vertex v in OPEN list that has minimal value of estimated time D, which is the sum of actual time from source vertex to it and its estimated time to destination (based on Euclidean distance and maximal allowed speed), and the shortest path from src to the expanded vertex v is recorded. We continue doing the search process (i.e., the expansion) until one intermediate destination is finally reached.

Due to the high dynamics of road condition in rash hour, it is essential to monitor the road condition and react to the relevant updates on it. In the IRS strategy, we conduct road condition monitoring in the same way as [9]. On arrival of a batch of delay updates, we check if the number of road segments affected by the updates that lie inside the pre-computed ellipse for a routing query exceeds ε (a threshold) times the average number of segments lying inside an ellipse of this area. If so, we re-run the intermediate destination and route search and return a real-time optimal path to the end user. In this way, computational overhead for the continuous monitoring can be significantly reduced.

11.4   Summary

Traffic navigation is a basic service for people's travel nowadays. However, most studies focus on route planning on static road network, without considering the high dynamics of road network, which can greatly affect the performance of route search. In this chapter, we present a traffic aware route planning strategy based on incremental planning method. By selecting intermediate destinations, a partial path rather than whole path is planned each time for long distance queries. In this way, route planning is more efficient because it is carried out in a much smaller region, and unnecessary re-calculations caused by the dynamic road conditions can be avoided.

References

1. Bast H, Funke S, Matijevic D, Sanders P, Schultes D (2007) In transit to constant time shortest-path queries in road networks. In: Proceedings of the workshop on algorithm engineering and experiments (ALENEX 2007), New Orleans
2. Chen S, Tu YC, Xia Y (2011) Performance analysis of a dual-tree algorithm for computing spatial distance histograms. VLDB J 20(4):471–494
3. Demiyurek U, Kashani FB, Shahabi C, Ranganathan A (2011) Online computation of fastest path in time-dependent spatial networks. In: Proceedings of 12th international symposium on spatial and temporal databases (SSTD 2011), Minneapolis, pp 92–111
4. Deng K, Zhou X, Shen HT (2007) Multi-source skyline query processing in road networks. In: Proceedings of the 23rd international conference on data engineering (ICDE 2007), Istanbul, pp 796–805

5. Ding B, Yu JX, Qin L (2008) Finding time-dependent shortest paths over large graphs. In: Proceedings of the 11th international conference on extending database technology (EDBT 2008), Nantes, pp 205–216
6. Gao J, Jin R, Zhou J, Yu JX, Jiang X, Wang T (2011) Relational approach for shortest path discovery over large graphs. Proc PVLDB 5(4):358–369
7. Gao J, Qiu H, Jiang X, Wang T, Yang D (2010) Fast top-k simple shortest paths discovery in graphs. In: Proceedings of the 19th ACM conference on information and knowledge management (CIKM 2010), Toronto, pp 509–518
8. Geisberger R, Sanders P, Schultes D, Delling D (2008) Faster and simpler hierarchical routing in road networks. In: Proceedings of 7th international workshop WEA 2008, Provincetown, pp 319–333
9. Goldberg A, Harrelson C (2005) Computing the shortest path, a search meets graph theory. In: Proceedings of the 6th annual ACM-SIAM symposium on discrete algorithms (SODA 2005), Vancouver, pp 156–165
10. Gonzalez H, Han J, Li X, Myslinska M, Sondag J (2007) Adaptive fastest path computation on a road network: a traffic mining approach. In: Proceedings of the 33rd international conference on very large data bases (VLDB 2007), Vienna, pp 794–805
11. Gunturi V, Nunes E, Yang K, Shekhar S (2011) A critical-time-point approach to all-start-time Lagrangian shortest paths: a summary of results. In: Proceedings of 12th international symposium on spatial and temporal databases (SSTD 2011), Minneapolis, pp 74–91
12. Hua M, Pei J (2010) Probabilistic path queries in road networks: traffic uncertainty aware path selection. In: Proceedings of the 13th international conference on extending database technology (EDBT 2010), Lausanne, pp 347–358
13. Huang B, Wu Q, Zhan F (2007) A shortest path algorithm with novel heuristics for dynamic transportation networks. Int J Geogr Inf Sci 21(6):625–644
14. Kanoulas E, Du Y, Xia T, Zhang D (2006) Finding fastest paths on a road network with speed patterns. In: Proceedings of the 22nd international conference on data engineering (ICDE 2006), Atlanta, p 10
15. Koenig S, Likhachev M, Furcy D (2004) Lifelong planning A*. Artif Intell 155(1–2):93–146
16. Kriegel H, Renz M, Schubert M (2010) Route skyline queries: a multi-preference path planning approach. In: Proceedings of the 26th international conference on data engineering (ICDE 2010), Long Beach, pp 261–272
17. Li F, Chen D, Hadjieleftheriou M, Kollios G, Teng S (2005) On trip planning queries in spatial databases. In: Proceedings of 7th international symposium on spatial and temporal databases (SSTD 2005), Angra dos Reis, pp 273–290
18. Likhachev M, Ferguson D, Gordon G, Stentz A, Thrun S (2005) Anytime dynamic A*: an anytime, replanning algorithm. In: Proceedings of ICAPS 2005, Monterey, pp 262–271
19. Malviya N, Madden S, Bhattacharya A (2011) A continuous query system for dynamic route planning. In: Proceedings of the 27th international conference on data engineering (ICDE 2011), Hannover, pp 792–803
20. Mouratidis K, Yiu M, Papadias D, Mamoulis N (2006) Continuous nearest neighbor monitoring in road networks. In: Proceedings of the 32nd international conference on very large data bases (VLDB 2006), Seoul, pp 43–54
21. Potamias M, Bonchi F, Castillo C, Gionis A (2009) Fast shortest path distance estimation in large networks. In: Proceedings of the 18th ACM conference on information and knowledge management (CIKM 2009), Hong Kong, pp 867–876
22. Rice M, Tsotras V (2010) Graph indexing of road networks for shortest path queries with label restrictions. Proc PVLDB 4(2):69–80
23. Samet H, Sankaranarayanan J, Alborzi H (2008) Scalable network distance browsing in spatial databases. In: Proceedings of the ACM SIGMOD international conference on management of data (SIGMOD 2008), Vancouver, pp 43–54
24. Sanders P, Schultes D (2005) Highway hierarchies hasten exact shortest path queries. In: Proceedings of ESA 2005, Palma de Mallorca, pp 568–579

25. Sankaranarayanan J, Samet H, Alborzi H (2009) Path oracles for spatial networks. Proc PVLDB 2(1):1210–1221
26. Stentz A (1994) Optimal and efficient path planning for partially-known environments. In: Proceedings of the international conference on robotics and automation, San Diego, pp 3310–3317
27. Xiao Y, Wu W, Pei J, Wang W, He Z (2009) Efficiently indexing shortest paths by exploiting symmetry in graphs. In: Proceedings of the 12th international conference on extending database technology (EDBT 2009), Saint-Petersburg, pp 493–504
28. Xu J, Guo L, Ding Z, Sun X, Liu C (2012) Traffic aware route planning in dynamic road networks. In: Proceedings of the 17th international conference on database systems for advanced applications (DASFAA 2012), Busan, pp 576–591

Chapter 12
Location Privacy

Abstract With rapid development of sensor and wireless mobile devices, it is easy to access mobile users' location information anytime and anywhere. On one hand, LBS is becoming more and more valuable and important. On the other hand, location privacy issues raised by such applications have also gained more attention. However, due to the specificity of location information, traditional privacy-preserving techniques in data publishing cannot be used. In this chapter, we will introduce location privacy, analyze the challenges of location privacy preserving, and give a survey of existing work including the system architecture, location anonymity, and query processing.

Keywords Location-based service • Moving object • Location privacy • Privacy preserving • Location anonymization

12.1 Introduction

In LBS applications, mobile users send their location information to service providers and enjoy various types of location-based services, such as mobile yellow page (e.g., "Where is my nearest restaurant"), mobile buddy list (e.g., "Where is my nearest friend"), traffic navigation (e.g., "What is my shortest path to the Summer Palace"), and emergency support services (e.g., "I need help and send me the nearest police"). LBS is playing an important role in people's daily life. In 2010, the total population of GPS-enabled LBSs subscribers reached 315 million, up from 12 million in 2006, according to a new study from ABI Research.

While people get much benefit from the useful and convenient information provided by LBSs, the privacy threat of revealing a mobile user's personal information (including the identifier and location) has become a severe issue. It has been

X. Meng et al., *Moving Objects Management: Models, Techniques and Applications*, DOI 10.1007/978-3-642-38276-5_12,
© Tsinghua University Press, Beijing and Springer-Verlag Berlin Heidelberg 2014

reported in [17] and an article in USA Today Dec., 2002[1] that a man was tracking his ex-girlfriend with GPS. Some companies increasingly use GPS-enabled cell phones to track employees.[2] Although many cases illustrate the various benefits of mobile devices, the user's privacy is threatened. In order to enjoy a good quality of LBS, an exact location is needed. However, an exact location needs to be hidden, to meet privacy requirements.

In this chapter, we first give an LBS example and show what kind of privacy issues are threaten. Then, the state of the art of privacy in location-based services is introduced, including challenges, system architecture, and cloaking methods. Finally, compared with traditional query processing, the key challenges of privacy-aware query on moving objects are introduced.

12.2   Privacy Threats in LBS

A major privacy threat specific to LBS usage is the location privacy breaches [8]. Such breaches take place when a party that is not trusted gets access to information that reveals the locations visited by the individual as well as the times during which these visits took place. An adversary can utilize such location information to infer details about the private life of an individual, such as political affiliations, alternative lifestyles, or medical problems of an individual, or the private businesses of an organization, such as new business initiatives and partnerships. First, using her PDA phone, Alice issues a query to the service provider (e.g., Google Map) to find out "where is the nearest hospital with specialty in cancer." Alice wants to hide her exact location (e.g., being in a hospital or at home), as well as the information that it is her (Alice) who issued a query about cancer. Or else, an adversary may infer that Alice has some medical problem.

Location privacy is a particular type of information privacy [2]. Westin defined **information privacy** as "the claim of individuals, groups, or institutions to determine for themselves when, how, and to what extent information about them is communicated to others" [21]. Whereas **location privacy** is defined as the ability to prevent other parties from learning one's current or past location. **Sensitive data** [3] refers to information of general concern, like medical information or financial data that could be transmitted as part of a service request; it may also be the spatio-temporal information regarding the user, as possibly collected by a location-based service provider. Examples include (1) information on the specific location of

---

[1]"GPS system used to stalk woman" (http://www.usatoday.com/tech/news/2002-12-30-gps-stalker_x.html).

[2]"Companies increasingly use GPS-enabled cell phones to track employees" (http://wifi.weblogsinc.com/2004/09/24/companies-increasingly-use-gps-enabled-cell-phones-to-track/) Weblogsinc. September, 2004

individuals at specific times, (2) movement patterns of individuals (specific routes at specific times and their frequency), and (3) personal points of interest (frequent visits to specific shops, clubs, or institutions).

Privacy threats related to location-based services are classified into two categories [11]: communication privacy threats and location privacy threats. In the communication privacy domain, sender anonymity is maintained, which implies that eavesdroppers on the network and LBS providers cannot determine the originator of a message. Compared to non-LBS web services, the location information is the key problem: an adversary can re-identify the sender of an otherwise anonymous message by correlating the location information with prior knowledge or observations about a subject's location. Consider the case where a subject reveals his/her location L in a message M to a location-based service and an adversary A has access to this information. Then, sender anonymity and location privacy is threatened by location information in the following ways:

- *Restricted Space Identification.* If A knows that space L exclusively belongs to subject S then A learns that S is in L and S has sent M. For example, when the owner of a suburban house sends a message from his garage or driveway, the coordinates can be correlated with a database of geocoded postal addresses to identify the residence. An address lookup in phone or property listings then reveals the owner and likely originator of the message.
- *Observation Identification.* If A has observed the current location L of subject S and finds a message M from L, then A learns that S has sent M. For example, the subject has revealed its identity and location in a previous message and then wants to send an anonymous message. The latter message can be linked to the previous one through the location information.
- *Location Tracking.* If A has identified subject S at location L_i and can link a series of location updates $L_1, L_2, \ldots, L_i, \ldots, L_n$ to the subject, then A learns that S visited all locations in the series.

Location privacy threats describe the risk that an adversary learns the locations that a subject visited (and the corresponding times). Through these locations, the adversary receives clues about private information such as political affiliations, alternative lifestyles, or medical problems. Assuming that a subject does not disclose his/her identity at such a private location, an adversary could still gain this information through location tracking. If the subject transmits his/her location with high frequency, the adversary can, at least in less populated areas, link subsequent location updates to the same subject. If at any point the subject is identified, his/her complete movements are also known.

There have been a number of follow-up studies based on location privacy preserving, which can be divided into two directions.

1. How to perform location anonymization. **Anonymity** is the state of being not identifiable within a set of subjects, referred to the anonymity set [19]. Location anonymity guarantees the inability to associate location information to a particular individual/group/institution through inference attacks [15]. Specifically, its

goal is to prevent disclosure of unnecessary information, including the individual identity and location of an individual, through explicit or implicit control of what information is given to "whom and when."

2. How to efficiently answer location-based queries (e.g., nearest neighbor and range queries) with cloaked regions [12, 18]. In a privacy-aware LBS system, location information is fuzzy instead of being exact. It can be a set of locations or an obfuscated location. Such that query processing in traditional moving object databases is not applicable now. We have to extend it or find new methods for answering queries with anonymized location.

The challenges faced in location privacy preserving can be summarized as follows:

1. It needs a trade-off between location privacy protecting and location-based services enjoying. As the data precision increases, so does the data utility; however, the privacy is threatened. It is often desirable to strike a balance between the location privacy and quality of services (QoS) requirements. A similar case occurs with regard to location privacy preserving. When a user issues a query, he has to publish his exact location. The more exact the location data is, the QoS correspondingly rises, but the privacy preserving is at a very low level. The QoS here includes response time, communication cost, etc.

2. Location information is multidimensional data, and they are dependent with each other. By contrast, data in publishing has independent attributes. And the attribute has one-dimensional value. In privacy preserving in data publishing, the data are partitioned into different groups based on all attributes. The anonymization method on each dimension can be different. However, in location privacy preserving, location is multidimensional information. We cannot handle it separately.

3. Location privacy preservation is online and service centric, which should tolerate the high frequency of location updates. Data anonymization in data publishing is applicable for the current snapshot of data. It is off-line and data centric. Therefore, it has no constraints on response time. However, for location privacy protection, the processor has to face so many moving objects with locations being updated frequently. Therefore, the cloaking time is a very important factor for location anonymization. Meanwhile, the problem of privacy compromise in location cloaking for continuous location updates should be considered, e.g., trajectory anonymization.

4. QoS is a very important factor. In privacy protection in data publishing, the focus is only on whether the user's privacy information is protected. However, in location privacy preserving, privacy protection is only one of the several issues. Other issues include how long users have to wait for the query answer and how much it costs when the answer is got. Specifically, the location is fuzzy as a result of anonymization. Therefore, it is a challenge to provide highly efficient, accurate, and anonymous location-based services based on the knowledge of the cloaked spatial areas rather than the exact location information. Therefore, how to provide highly efficient, accurate, and anonymous location-based services

based on the knowledge of the cloaked spatial areas rather than the exact location information is another problem that users are concerned.

5. Privacy requirements are personalized. Different people have different privacy requirements. Moreover, the privacy levels for the same person may be different when the place or time is different. For example, when someone is shopping, his/her privacy level is low. However, if he/she is in a hospital, the privacy level increases. Therefore, we cannot unify everyone's privacy requirements or force users to accept a minimum level of privacy.

In order to accommodate personalized privacy requirements, each user can specify four parameters for protecting the location privacy at least:

- k: It represents the anonymity level in the location k-anonymity model. More specifically, each cloaked region should cover at least k different users. The larger the value of k, the more privacy is protected.
- A_{min}: It specifies the minimum area that the cloaked region should have. This is to prevent the cloaked region from being too small for highly populated areas.
- A_{max}: It constrains the maximum area of the cloaked region. As the area of the cloaked region would affect the accuracy and size of the query result, this parameter stands for one kind of quality of service.
- δ_t: It is the maximum tolerable cloaking delay, which is a QoS parameter. The larger is the δ_t value, the worse is the service quality, since the user will have a higher chance of moving away from the location where the query was issued.

The former two parameters are the constraints for location anonymization, which is the minimum of QoS. And the latter two are constraints for location service quality, which indicate the worst QoS.

12.3 System Architecture

System architectures for location privacy are classified into three categories: non-cooperative architecture, centralized architecture, and peer-to-peer architecture. Users in non-cooperative architecture depend only on their knowledge to preserve their location privacy. However, in centralized architecture, a centralized entity is responsible for gathering information and providing the required privacy for each user. For peer-to-peer architecture, users collaborate with each other without the centralized entity to provide customized privacy for each single user.

12.3.1 Non-cooperative Architecture

The non-cooperative architecture system [4] consists of many mobile users and an untrusted service provider. It is assumed in this architecture that each of the clients is location-aware – they can position their own locations (e.g., using GPS or WLAN based positioning). It has strong capability for calculation and storage to get the anonymized location according to the personalized privacy requirement.

Location obfuscation is performed at client's end. On receiving the anonymized location, the untrusted service provider processes the request and sends back the candidate results to the user. As the client knows its own exact location, it obtains the true result on its own. In a word, the location anonymization and results refinement are both completed by clients themselves.

The good point of this architecture is that it is simple and easy to be incorporated with other technologies. But the requirement for client is too high. The most worst is that it generates the anonymized location only by its own knowledge, but ignores the other users' locations. Therefore, privacy is easily threatened in this architecture. For example, [6] reduces the resolution of location for location privacy protection, and thus a cloaked region is issued. However, only one user is covered in this region, such that the query issued from this region can be easily to be matched with the issuer. Query privacy is disclosed.

12.3.2 Centralized Architecture

The system consists of many mobile users, a trusted anonymizing proxy, and an untrusted service provider. Compared with non-cooperative architecture, a third-party anonymization proxy (middleware) is required for all communications between mobile users and LBS applications. Its functions can be summarized as follows:

- It receives the exact locations from clients.
- It blurs the locations and sends the blurred locations to the service provider.
- It receives and refines the candidate results, which are sent by the service provider. Moreover, it relays the exact query result to clients.

Mobile clients communicate with third-party LBS providers through the anonymity proxy. The mobile user sends location-based queries to the anonymizing proxy. The anonymity proxy is a secure gateway to the LBS providers for the mobile clients. Upon receiving the location-based query, the anonymizing proxy removes any identifiers, such as IP addresses. In the meantime, it invokes the location cloaking algorithm to generate a cloaked region in accordance with the user's privacy requirement. Then, it forwards the modified query to the service provider. Finally, the anonymizing proxy will relay the result returned from the service provider to the mobile user.

With a trusted anonymizing proxy, it provides powerful privacy guarantees with high-quality services. But it still suffers from that [10]:

- The centralized anonymizer proxy is a bottleneck due to handling of query requests, frequent updates of user locations, and result post-processing. Moreover, the anonymizer is a single point of failure; the system cannot function without it.

- The complete knowledge of the locations and queries of all users is a serious security threat, if the anonymizer is compromised. Even if there is no attack, the centralized anonymizer may be subject to governmental control and may be banned or forced to disclose sensitive user information.

12.3.3   Peer-to-Peer Architecture

Similar to non-cooperative architecture, peer-to-peer architecture consists of mobile clients and service providers. However, the users collaborate with each other to keep their customized privacy information. In this aspect, peer-to-peer architecture is different from non-cooperative architecture.

Each mobile user carries mobile devices (e.g., mobile phones, PDAs) with embedded positioning capabilities (e.g., GPS). The devices have processing power and access the network through a wireless protocol such as WiFi, GPRS, or 3G. Moreover, each device has a unique network identity (e.g., IP address) and can establish point-to-point communication (e.g., TCP/IP sockets) with any other devices in the system through a base station (i.e., the two devices do not need to be within communication range of each other). For security reasons, all communication links are encrypted.

In addition, there is a trusted central Certification Server (CS), where users are registered. Prior to entering the system, a user u must authenticate against the CS and obtain a certificate. Users having a certificate are trusted by all other users. Typically, a certificate is valid for a few hours; it can be renewed by recontacting the CS. Apart from the certificate, the CS returns to u the IP addresses of some users who are currently in the system. u uses this list to identify an entry point to the distributed network. Note that the CS does not know the locations of the users and does not participate in the anonymization process. Therefore, the workload of the CS is low (i.e., no location updates); moreover, it does not store any sensitive information.

Each user corresponds to a peer. Peers are partitioned into different groups, according to their location. Within each group, peers elect a head. The anonymization process can be completed by the group head or the user who issues the service. However, the group head refines the candidate results for the users in its group. To achieve load balancing, group heads can be rotated in a round-robin manner [10]. There are three main issues to be addressed in this architecture: anonymization, query processing, and head selection. Group Formation [5] and PRIVE [10] are the two representative works.

12.4   Location Anonymization Techniques

The goal of location anonymization is to protect the user's location while meeting user-specified QoS requirements. A query in LBS can be formalized as $r = (id, l, q)$, where id is the user's identifier, $l = (x, y)$ is the user's current location,

Fig. 12.1 Location
4-anonymity

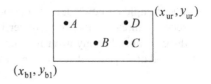

(x_{ur}, y_{ur})

$\bullet A$ $\bullet D$

$\bullet B$ $\bullet C$

(x_{bl}, y_{bl})

and q is the query content. These three parameters have different implications.
First, id uniquely identifies a user. It cannot be revealed to any third party and
should be removed before being forwarded to the LBS server. Second, l could
be a quasi-identifier (QI) attribute, which cannot directly identify a user but may
reveal a user's association with requests by joining with external data (e.g., some
background knowledge such as yellow pages and location data obtained by network-
based positioning). Thus, l should be cloaked (enlarged) in the request sent to
the LBS server. Third, q is a sensitive attribute, which may be confidential to
an individual (subject to her/his preference) but must be sent to the LBS server
in order to answer the request. Following the above analysis, the simplest way
is to replace his/her identity with a pseudonym before sending the query to the
service provider. However, as described in Sect. 12.2, it is not enough. We have to
anonymize the location information. There have been a number of follow-up studies
on this issue [4, 6, 10, 13–15, 22].

12.4.1   Location K-Anonymity Model

Location k-anonymity model is the most widely accepted metric for location privacy
preserving. The k-anonymity model was originally proposed for privacy protection
in data publishing by Sweeny [20]. As defined in [20], a release of data provides
k-anonymity protection if the information for each individual contained in the
release cannot be distinguished from at least $k - 1$ individuals whose information
also appear in the release.

To address the location privacy issue, location k-anonymity was proposed by
Gruteser and Grunwald [11]. A mobile user is considered as location k-anonymous
if and only if the location information sent to the service provider is indistinguish-
able from those of at least $k - 1$ other users. More specifically, location information
is represented by a tuple containing three intervals $([x_1, x_2], [y_1, y_2], [t_1, t_2])$. The
intervals $[x_1, x_2]$ and $[y_1, y_2]$ describe a two-dimensional area where the subject
is located. $[t_1, t_2]$ describes a time period during which the subject was present in
the area. Note that the intervals represent uncertainty ranges; we only know that at
some point in time within the temporal interval, the subject was present at some
point of the area given by the spatial intervals. Thus, a location tuple for a subject
is k-anonymous, when it describes not only the location of the subject but also
the locations of $(k - 1)$ other subjects. In other words, $(k - 1)$ other subjects
also must have been presented in the area and the time period described by the
tuple. For example, Fig. 12.1 shows a location 3-anonymity example (for stating

Table 12.1 Location 4-anonymity

User	Real location	Anonymity location
A	(x_A, y_A)	$([x_{bl}, x_{ur}], [y_{bl}, y_{ur}])$
B	(x_B, y_B)	$([x_{bl}, x_{ur}], [y_{bl}, y_{ur}])$
C	(x_C, y_C)	$([x_{bl}, x_{ur}], [y_{bl}, y_{ur}])$
D	(x_D, y_D)	$([x_{bl}, x_{ur}], [y_{bl}, y_{ur}])$

conveniently, time interval is omitted here). Locations of A, B, C, and D are all extended to a rectangle $CR = ([x_{bl}, x_{ur}], [y_{bl}, y_{ur}])$, where (x_{bl}, y_{bl}) and (x_{ur}, y_{ur}) are the bottom-left and up-right location of cloaked region. If it is represented by a table form, it is shown as Table 12.1. Thus, the adversary cannot be sure the exactly location of each mobile user. The users in the cloaked region constitute the cloaking set. In this example, cloaking set is $\{A, B, C, D\}$. Generally speaking, the larger the anonymity set k is, the higher is the degree of anonymity. Note here that k is specified by the user, which is one of the four parameters mentioned in Sect. 12.2. Generally speaking, the larger k is, the larger the size of cloaked region is. It largely depends on the surrounding environment. Let $k = 100$, and if the user is in the shopping mall, the cloaked region will be very small. However, if the user is in the desert, the cloaked region may be very large.

12.4.2 p-Sensitivity Model

Several methods have been proposed to support location-based services without revealing mobile users' privacy information. There are two types of privacy concerns in location-based services: location privacy and query privacy. Existing studies, based on location k-anonymity, mainly focus on location privacy and are insufficient to protect query privacy. In particular, due to lack of semantics, location k-anonymity has the drawback of query homogeneity attack. In many LBS applications, mobile users do not mind to reveal their exact location information. However, they would like to hide the fact that they have issued queries that contain sensitive content as such information may reveal their personal interest (e.g., searching the nearest clinic when the user is in an insensitive public place). In this section, we will discuss protection of query privacy for LBS applications.

Existing location k-anonymity technique can be used to improve protection of query privacy. Nevertheless, the protection provided by location k-anonymity is not sufficient. Consider a scenario where each query location is enlarged in accordance with k-anonymity. That is, each query location is covered by at least k queries (hereafter called anonymity set). Thus, even though the adversary knows the exact location of a user, he is not able to link the user to a specific query (rather k queries). However, one main weakness of k-anonymity is that it considers only spatial proximity in forming anonymity sets, but not query semantics. In an extreme case, if all queries in the anonymity set contain the same content, the query privacy is still revealed. This situation is not uncommon. For example, when friends meet

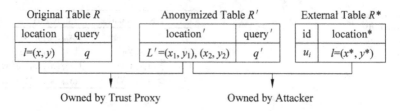

Owned by Trust Proxy   Owned by Attacker

Fig. 12.2 Original table, anonymized table, and external table

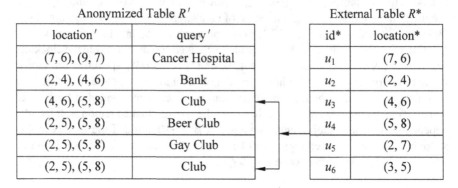

Anonymized Table R'		External Table $R*$	
location$'$	query$'$	id*	location*
(7, 6), (9, 7)	Cancer Hospital	u_1	(7, 6)
(2, 4), (4, 6)	Bank	u_2	(2, 4)
(4, 6), (5, 8)	Club	u_3	(4, 6)
(2, 5), (5, 8)	Beer Club	u_4	(5, 8)
(2, 5), (5, 8)	Gay Club	u_5	(2, 7)
(2, 5), (5, 8)	Club	u_6	(3, 5)

Fig. 12.3 Query homogeneity attack

after office hours and discuss visiting some club, they may all issue location-based queries containing the keyword "club." Since these query locations are spatially proximate, they are very likely to be anonymized together in the same anonymity set. As a result, although the adversary cannot infer which user issued which query, he would know all users queried about clubs. Consider another example, several specialty clinics are located in a small area of the downtown, and people would easily lose their way after leaving the highway exit. The users may often issue a location-based query to find the way to some specialty clinic near the highway exit. These queries are then likely to be anonymized with each other. Furthermore, even if the k queries in an anonymity set are not of the same kind (e.g., satisfying l-diversity in [16]), it is still not acceptable to some users if they all contain sensitive information (e.g., some queries ask about clubs and some others ask about clinics). In a word, due to lack of semantics, location k-anonymity can just prevent the association between users and requests, but not the association between users and (sensitive) query contents, and hence suffers from the aforementioned attacks.

To protect query privacy, first we define the query semantics. For simplicity, we simply assume that each query can be classified into two types according to its content: (1) insensitive query (Q_i), e.g., queries about traffic, and (2) sensitive query (Q_s), e.g., queries about bar, clinic, and political information.

Following our assumption, the attacker may obtain the tables R' and $R*$ (as shown in Fig. 12.2) and attempt to establish their relationship. We use an example to illustrate each of these two attacks. Figure 12.3 shows an example of query homogeneity attack. We assume that there are six users u_1 through u_6. In the external table R^*, user u_4 has the location of $l_4^* = (5, 8)$. When l_4^* is joined with R', the attacker can observe that l_4^* is covered by the cloaking regions of four requests, each of which covers more than one location in R^*. Thus, the attacker can only know that u_4 has sent one of the four queries but cannot tell which one. However, all the four queries are about "club." Hence, the attacker can conclude that u_4 must have queried about "club," which might be sensitive with respect to u_4's privacy preference. In a word, the attacker can infer that a user has issued some sensitive query with high confidence.

To protect against location linking attack, each query can be de-linked from its issuer by confusing the attacker with more than one users appearing in the cloaking region of the query; each user can be de-linked from his/her query by confusing the attacker with more than one query having cloaking regions that cover the user's location.

Given an anonymized query r', denote by $P(r' \rightarrow u^*)$ the probability of the user u^* in r', S_u being the true issuer of r'. Given a user u^*, denote by $P(u^* \rightarrow r')$ the probability of the query r' in u^*, S_r being sent by u^*. By the assumption of uniform background knowledge, the probability of a query being sent by any user in its S_u is equal and each user has the same probability of sending any request in the user's S_r. Thus, in order to defend against location linking attack, it is required that $P(r' \rightarrow u^*)$ and $P(u^* \rightarrow r')$ are both less than or equal to the user-specified threshold $\frac{1}{k}$:

$$P(r' \rightarrow u^*) = \frac{1}{|r'.S_u|} \leq \frac{1}{k} \tag{12.1}$$

$$P(u^* \rightarrow r') = \frac{1}{|u^*.S_r|} \leq \frac{1}{k} \tag{12.2}$$

To protect against query homogeneity attack, each user can be de-linked from sensitive queries by confusing the attacker with some insensitive queries in the user's S_r. Given a user u^*, denote by $P(u^* \rightarrow Q_s)$ the probability that u^* has sent some sensitive query. Hence, it is required that $P(u^* \rightarrow Q_s)$ is always less than the user-specified threshold p. It can be formalized as

$$P(u^* \rightarrow Q_s) = \frac{\sum_{r_i \in u^*.S_r} v_i}{|u^*.S_r|} < p \tag{12.3}$$

where v_i is the sensitivity value of query r_i and Σv_i computes the total number of sensitive queries in the request anonymity set of user u^*.

Equations (12.1) and (12.2) ensure that any query will be linked with at least k users and any user will be linked with at least k queries. Equation (12.3) ensures that the probability of any user sending some sensitive query is less than p. Finally, we wrap up the p-sensitivity model as

Fig. 12.4 Dummies

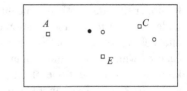

Fig. 12.5 Cloaking

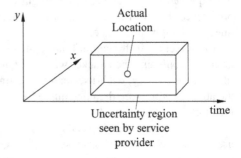

***p*-Sensitivity**: *p*-sensitivity is satisfied if and only if:

- For each user u^*, $P(u^* \to r') \leq \frac{1}{k}$, $P(u^* \to Q_s) < p$.
- For each query r', $P(r' \to u^*) \leq \frac{1}{k}$.

12.4.3 Anonymization Algorithms

In terms of the techniques used for protecting location privacy, the existing approaches can be classified into three categories: dummy, cloaking, and encryption.

The first technique is to generate dummies. A user specifies a dummy location instead of his/her genuine location. As shown in Fig. 12.4, circle point represents the query, and the square point represents the object queried. The black point represents the true location, and the white points represent dummies. The user location is represented with a wrong value, such that the privacy is achieved from the fact that the reported location is false. The QoS and the amount of privacy mainly depend on how far the dummy is from. The larger the distance, the worst QoS, but much privacy is preserved.

The second technique is cloaking. The main idea of cloaking is to reduce the spatio-temporal resolution of the user location. A precise location is replaced with a cloaked region, which is shown in Fig. 12.5, so that the attacker cannot know the exact location of the user. The cloaked region is a closed region, which can be any shape with a predefined probability distribution of this object in the region. In general, most existing work uses a rectangle or a circle to present a cloaked region and assumes that the probabilities of the users being in a cloaking region are the same. The difference between cloaking and dummy is that the location in the former case is a fuzzy location, whereas in the latter case, the locations are all precise and the attacker just cannot tell which one is real. The larger is the cloaked region, the more privacy is preserved, but the less specific is the request.

Third, some work [9] suggested using encryption for location privacy protection recently. Its main idea is that the query is encrypted so that the service provider answers the queries without knowing what kind of information is being retrieved. Then, the user de-encrypts the result candidates and refines them at the client side. For example, Ghinita et al. [9] proposed a framework that is based on Private Information Retrieval (PIR). The framework partitions the space into grid cells and then the user requests the content of cell where he/she is located. Thanks to PIR, the user can encrypt which cell is requested while receiving the correct content.

12.5 Evaluation Metrics

Compared with dummies and encryption, cloaking is the most widely used method [1,7,8,13]. In this section, we introduce several evaluation metrics for system-level control of the balance between privacy value and performance implication in terms of QoS. These metrics [8] can be used to evaluate the effectiveness and the efficiency of anonymization algorithms based on cloaking.

Success rate is an important measure for evaluating the effectiveness of the proposed location k-anonymity model. It can be defined over a set $S' \subset S$ of requests as the percentage of messages that are successfully anonymized, which is formally represented as:

$$SR = \frac{|S'|}{|S|} \tag{12.4}$$

where S' is the number of requests that have been anonymized successfully and S is the number of requests issued.

Relative anonymity level is a measure of the level of anonymity provided by the cloaking algorithm, normalized by the level of anonymity required by the messages. It is measured by k'/k, where k' is the number of users actually included in the cloaking region while k is the number that user required. Note that the relative anonymity level cannot go below 1. Higher relative anonymity levels mean that on the average messages are getting anonymized with larger k values than the user-specified minimum k-anonymity levels. In general, we prefer algorithms that can provide higher relative anonymity levels.

Relative spatial resolution is a measure of the spatial resolution provided by the cloaking algorithm, normalized by the minimum acceptable spatial resolution defined by the spatial tolerances. Higher relative spatial resolution values imply that anonymization is performed with smaller spatial cloaking regions relative to the constraint boxes specified.

Relative temporal resolution is a measure of the temporal resolution provided by the cloaking algorithm, normalized by the minimum acceptable temporal resolution defined by the temporal tolerances. Higher relative temporal resolution values imply that anonymization is performed with smaller temporal cloaking intervals and thus with smaller delays due to perturbation. Relative spatial and temporal resolutions cannot go below 1.

Message processing time is a measure of the running time performance of the anonymization algorithm. It is the period from when a request is received to when the request is successfully cloaked. It includes the cloaking time as well as the waiting time for cloaking. The message processing time may become a critical issue, if the computational power at hand is not enough to handle the incoming messages at a high rate.

Important measures of efficiency include relative anonymity level, relative temporal resolution, relative spatial resolution, and message processing time. The first three are measures related with quality of service, whereas the last one is a performance measure.

12.6 Summary

This chapter presents the definition, the models, and the techniques of location privacy preserving. It consists of four main components. First, we introduced location privacy threats and gave an overview of the state-of-the-art research. Second, we presented three system architectures for location privacy preserving. Third, we discussed the various location privacy models and techniques. Finally, we introduced several evaluation metrics for system-level control of the balance between privacy value and performance implication in terms of QoS.

In real life, several major privacy threats are occurring due to the use of location-detection devices. Therefore, location privacy is a major obstacle in the ubiquitous deployment of location-based services. Location privacy protection is a new developing field, and there are several open issues to be researched.

References

1. Bamba B, Liu L, Pesti P, Wang T (2008) Supporting anonymous location queries in mobile environments with PrivacyGrid. In: Proceedings of the 17th international conference on world wide web (WWW 2008), Beijing, pp 237–246
2. Beresford AR, Stajano F (2003) Location privacy in pervasive computing. IEEE Pervasive Comput 2(1):46–55
3. Bettini C, Wang XS, Jajodia S (2005) Protecting privacy against location-based personal identification. In: Proceedings of the VLDB workshop on secure data management (SDM 2005), Trondheim, pp 185–199

4. Cheng R, Zhang Y, Bertino E, Prabhakar S (2006) Preserving user location privacy in mobile data management infrastructures. In: Proceedings of the 6th workshop on privacy enhancing technologies (PET 2006), Cambridge, pp 393–412

5. Chow CY, Mokbel MF, Liu X (2006) A peer-to-peer spatial cloaking algorithm for anonymous location-based services. In: Proceedings of the 14th ACM international symposium on geographic information systems (GIS 2006), Arlington, pp 171–178

6. Du J, Xu J, Tang Z, Hu H (2007) iPDA: supporting privacy-preserving location-based mobile services. In: Proceedings of the 8th international conference on mobile data management (MDM 2007), Mannheim, pp 212–214

7. Gedik B, Liu L (2005) Location privacy in mobile systems: a personalized anonymization model. In: Proceedings of the 25th international conference on distributed computing systems (ICDCS 2005), Columbus, pp 620–629

8. Gedik B, Liu L (2008) Protecting location privacy with personalized k-anonymity: architecture and algorithms. IEEE Trans Mob Comput 7(1):1–18

9. Ghinita G, Kalnis P, Khoshgozaran A, Shahabi C, Tan K (2008) Private queries in location based services: anonymizers are not necessary. In: Proceedings of the ACM SIGMOD international conference on management of data (SIGMOD 2008), Vancouver, pp 121–132

10. Ghinita G, Kalnis P, Skiadopoulos S (2007) MobiHide: a mobile peer-to-peer system for anonymous location-based queries. In: Proceedings of the 10th symposium on spatial and temporal databases (SSTD 2007), Boston, pp 221–238

11. Gruteser M, Grunwald D (2003) Anonymous usage of location based services through spatial and temporal cloaking. In: Proceedings of the 1st international conference on mobile systems, applications, and services (MobiSys 2003), San Francisco, pp 31–42

12. Hu H, Lee D (2006) Range nearest-neighbor query. IEEE Trans Knowl Data Eng 18(1):78–91

13. Kalnis P, Ghinita G, Mouratidis K, Papadias D (2006) Preserving anonymity in location based services. Technical report TRB6/06, Department of Computer Science, National University of Singapore

14. Kido H, Yanagisawa Y, Satoh T (2005) An anonymous communication technique using dummies for location-based services. In: Proceedings of the 2005 IEEE international conference on pervasive services (ICPS 2005), Santorini, pp 88–97

15. Liu L (2007) From data privacy to location privacy: models and algorithms. In: Proceedings of the 33rd international conference on very large data bases (VLDB 2007), Vienna, pp 1429–1430

16. Machanavajjhala A, Gehrke J, Kifer D (2006) l-diversity: privacy beyond k-anonymity. In: Proceedings of the 22nd international conference on data engineering (ICDE 2006), Atlanta, pp 24–37

17. Man Accused of Stalking Ex-girlfriend with GPS. Fox News. September, 2004. http://www.foxnews.com/story/0,2933,131487,00.html

18. Mokbel MF, Chow C, Aref WG (2006) The New Casper: a privacy-aware location-based database server. In: Proceedings of the international conference on very large data bases (VLDB 2006), Seoul, pp 1499–1500

19. Pfitzmann A, Koehntopp M (2000) Anonymity, unobservability, and pseudonymity. A proposal for terminology. In: Proceedings of the international workshop on design issues in anonymity and unobservability, Berkeley, pp 1–9

20. Sweeney L (2002) K-anonymity: a model for protecting privacy. Int J Uncertain Fuzziness Knowl Based Syst 10(5):557–570

21. Westin AF (1967) Privacy and freedom. Atheneum, New York

22. Xiao Z, Meng X, Xu J (2007) Quality-aware privacy protection for location-based services. In: Proceedings of the 12th international conference on database systems for advanced applications (DASFAA 2007), Bangkok, pp 434–446

Index

X. Meng et al., *Moving Objects Management: Models, Techniques and Applications*, DOI 10.1007/978-3-642-38276-5,
© Tsinghua University Press, Beijing and Springer-Verlag Berlin Heidelberg 2014

inted in the United States
ookmasters